Lost Causes i

R.F. Streater

Lost Causes
in and beyond
Physics

Springer

Professor Dr. R.F. Streater
University of London
King's College London
Deptartment of Mathematics
WC2R 2LS London
United Kingdom
E-mail: raymond.streater@kcl.ac.uk

ISBN   978-3-642-07168-3   e-ISBN   978-3-540-36582-2

Springer is a part of Springer Science+Business Media
springer.com
© Springer-Verlag Berlin Heidelberg 2010

Cover design: *design & production* GmbH, Heidelberg

Preface

This book arose from a suggestion of the Editor, Dr. Christian Caron, that my web-site, having a similar title, might be published as a book.

My web-site started as an account of hidden variables, which had been promoted by David Bohm. It was claimed by Bohm that his theory was based on classical probability theory, and gave the same results as quantum theory. This seemed to be true, in that his theory led to the construction of a solution to density of probability for the position variable which is the same as that given by quantum mechanics, $|\psi(\mathbf{x}, t)|^2$, where ψ obeys the Schrödinger equation. Later, Bell obtained his inequalities; these showed that any theory based on classical (that is, random) observables with four variables led to correlations obeying a new inequality; this was tested in spin systems and found to be violated. The results were, however, very close to the predictions made by quantum theory. My conclusion to this was that quantum theory was correct, and classical theory incorrect in spin systems. Bohm however, persisted, and claimed that it meant that quantum theory was non-local. He had found an unwarrented assumption in Bell's proof of his inequalities: the assumption of locality. Bell had indeed assumed that in a random theory, there might be some "hidden" variables for each observer. When there are two observers, X and Y, Bell postulated that the observables of X were independent of any hidden variables of Y, and that the observables of Y were independent of any hidden variables of X. This was the extra assumption of locality made by Bell. However, Landau, and Tsirelson, were able to derive the Bell inequalities without any such locality postulate; they just assumed that all the four observables appearing in the inequalities were random variables on the same sample space. They did not postulated anything about hidden variables; there might or might not be any. This work was not noticed by some authors; for example, Markopoulou and Smolin, in their paper "Quantum Theory from Quantum Gravity", arXiv.org/gr-gc/0311059, written six years after Landau's paper, and ten years after Tsirelson's, say that they begin without quantum theory, but that the non-local links in the theory provide the non-locality needed to avoid Bell's proof. Unfortunately, this cannot generate correlations that violate the Bell inequalities if all the observables are random variables on some space.

I decided to write the first part of the web-site, Lost Causes in Physics, after I received an e-mail asking whether, in view of Bell's theorem, Wightman and I were going to rewrite our book, **PCT, Spin and Statistics and All That**, to allow for non-local fields. The answer was NO. The trouble with Bohm's claim, that his theory agrees with quantum theory, is that he never considers the correlations between position-variables at different times. In the quantum theory, these might commute with each other, and so can be simultaneously measured, but in general will violate the Bell inequalities. This cannot happen if the position-variables are random variables, as they are in Bohm's model. It was proposed to me that I should really have included a page on David Bohm somewhere on my site; but I found it very difficult. Perhaps this book should replace the missing page; he was very original, but mostly wrong.

I first learnt about Frieden's attempt, to derive all the laws of physics from information theory, from a colleague who had bought the book, **Physics from Fisher Information** at the bookstall at the airport. He read some of it on the plane, and decided it was not worth keeping; he therefore threw it away on his arrival. It must have been bad! I started worrying about the thesis of Roy Meadow, concerning the probability that an unexplained cot-death was murder, in a blog set up to discuss physics. After I had added the topic to my Lost Causes, I received invaluable messages from Lynne Wrennall and Michael Nott. Later, Dr. Alex Belton sent me some very useful copies of various publications. Thanks are due to them. Thanks are also due to Dr. Tony Barnard, who showed me the paradox in classical probability based on Bayesian methods.

I would like to thank Dr. Caron for all his help in bringing about the book.

R.F. Streater

Contents

1

Introduction

When Aspect, Dalibrand and Roger reported the results of their photon experiments, [10], David Bohm called a press conference and announced it to the world. I thought I had misheard the first words, which were that contrary to the expectations of most physicists, Bell's inequalities have been violated by the experimental results. I was sure that he had meant to say that most physicists had expected just this result. After all, photons were known to be governed by quantum mechanics, and the results of the experiment hit the quantum predictions right on the nose. I thought that no serious physicist believed that the data would be consistent with classical theory. I should remind the reader that, in deriving the Bell inequalities, Bell used classical probability. So here was Bohm, a hitherto influential physicist, author of a respectable book on the subject, in a muddle over elementary questions.

A similar situation had arisen a decade earlier, when H. Dingle was asked by his publisher, Oliver and Boyd, to approve the reprinting of his excellent little book on special relativity. Dingle had, since writing that book, got muddled over the twin "paradox", and wanted to disown his own book. The publishers did not agree, and Dingle eventually approved the second edition. His condition was that a new preface be added, in which Dingle advised the reader not to believe anything in the book. Dingle's colleagues, Hermann Bondi and H.C. McCrae, made some effort to explain where Dingle went wrong, culminating in Bondi's complete and elementary exposition of the twin "paradox"; it isn't one. We shall explain this in Chap. (12). Though Dingle fought bravely for his point of view, the "controversy" died down and faded from the literature. I thought that the same would happen over Bohm's misunderstanding, but I was wrong. The debate expanded, and more and more physicists gave credence to Bohm's point of view. This view settled down to the idea that quantum mechanics was not wrong, but is nonlocal. Bohm reckoned that he had put his finger on a hidden and unjustified assumption in Bell's derivation of the inequalities: Bell had assumed that the theory was "local". This, therefore, must be where the derivation fails. However, Bell's locality assumption involved some assumed hidden variables, and is not needed by later workers

[110]. They derived the Bell inequalities simply by assuming that all observables are random variables on the same sample space. This is what I mean by the classical assumption made by Bell. This proof is given in Chap. (8).

After Bohm had died, his interpretation of quantum mechanics as true but nonlocal became the subject of many papers and conferences. I was asked, in view of the work of Bell and Bohm, whether Wightman and I were about to rewrite our book on local quantum field theory, *PCT, Spin and Statistics, and All That*. The answer was NO. Indeed, the "locality" found by Bohm in Bell's proof was unrelated to the locality postulate of quantum field theory, which is postulated by Wightman to mean that space-like separated observable fields commute with each other. Rather Bell made the following assumptions. He assumes that the theory is a random theory, and that some of the random variables correspond to observables, and that others do not. The latter are then the hidden variables. When a system splits into two parts, A and B, which become far apart, Bell assumes that each part consists of its observables and its hidden variables. The locality condition adopted by Bell was that the value of any observable of A is functionally independent of the values taken by the hidden variables at B. Similarly, the value of any of B's observables is assumed to be functionally independent of the values of A's hidden variables. This is only assumed to hold when A and B are far apart. With this assumption, together with the assumption that A's and B's observables are all random variables on the same probability space, Bell constructed a proof of his inequality. However, in Landau's proof of Bell's inequality, no such assumption on locality is made; the only assumption is that all the observables are representable as random variables on the *same* sample space. The violation of Bell's inequality by the quantum-mechanical spin correlations in the Einstein–Podolski–Rosen experiment [10] has nothing to do with the violation of such locality, but follows from the noncommutative nature of the algebra in the quantum theory of spin. This view will be justified later on in this work.

Thus, the "nonlocality" identified by Bohm was a misnomer; "contextual" is now the word preferred by philosophers of science. This means that classical probability (as used by Bell) can be used to describe the statistical results of a run of trials in which the same (compatible set) of observables is measured; but, if another compatible set is measured on the same system, in the same state, we must use a different classical statistical model: the model to be used depends on the (experimental) *context*. This is close to what Bohr was saying, forty years earlier. There is no one statistical model which describes all choices of experiment. This contextuality is not in general related to nonlocality in the physical sense: there is no propagation of material or information faster than light, and the measurement of one observable from a compatible set (which commute) does not disturb the statistics of any other observable from the set. Thus, Bohm's contextuality (called nonlocality by him) does not impact on relativistic causality or local commutativity in quantum field theory. Contextuality arises when we describe the quantum results using classical tools; there is no contextuality in the quantum description; the state (density operator) and the observables (self-adjoint operators) are the same whatever maximal abelian set is being described. Simply, the state is not a measure and the observables are not random variables on any one space. They are really there, though, in the mathematical sense; we do not support the idea that physics is not about real things.

In spite of this clear-cut answer to Bohmists, I continued to receive e-mails from students around the world, asking about the violation of locality. I therefore set up a web page setting out the Copenhagen point of view, trying to make it clear that there was nothing revolutionary in contextuality. I also included some other failed ideas, whose motivation might have been to avoid this perceived nonlocality or other paradoxes. Further unpromising work was done even trying to make use of the nonlocality, as in Stapp's theory of the brain. I realised that the entrenched belief in nonlocality was caused by a common error in statistics: many physicists jumped to the conclusion that a hundred percent correlation between two observations in quantum statistics indicates that the measurement done first *causes* the result of the other to happen. This error is well known in medical statistics (and is avoided by professionals). I therefore added a page, Good Cures, with a few topical remarks about medicine, where a little mathematics goes a long way. I thus describe the paradox of Einstein, Podolski and Rosen; this is shortened to the *EPR paradox*. Some of the ideas found in the paradox also arise in the study of the association between cannabis and schizophrenia. There is also a relation between the error made by those denying the collapse of the wave packet in quantum theory, and the error behind the work involving the assessment of the likelihood of guilt in multiple cot deaths, known as the Prosecutor's Fallacy. Both subjects need the concept of information and conditional probability.

There is no other elementary mathematical topic that is so badly taught to physicists as probability theory. I therefore start things off with a simple account of the misuse of the theory in the assessment of the probability that a pair of cot-deaths in the same family is due to murder. This is followed by an account of classical probability, including information theory and conditioning. The exposition is designed so that it is evident that classical probability is a special case of quantum probability, and Bayes's rule is a special case of the collapse of the wave packet. The subjective nature of probability emerges from the idea that information is, or is not, to hand. We need to develop elementary probability theory to the case when the same system is observed by several observers, a case not usually treated in the books. This leads to the ideas of game theory (but not into what is known as Bayesian probability). It is then easy to see that the *EPR* paradox is no such thing.

The same situation, with several observers in quantum theory, leads me to develop "the quantum game" in the chapter on quantum probability. This resolves those quantum paradoxes that arise because there is more than one observer; the trouble arises when the physicist does not distinguish the information available only to one observer from information available to the others. The upshot is that *EPR* is solved, there is no action at a distance, nor faster-than-light particles, nor retrospective changes of history. This idea removes the motivation of H. R. Stapp for his use of quantum theory in brain dynamics. The "weirdness" of quantum mechanics remains in the existence of entangled states (this is weird if you believe only in classical physics). Entangled states have the property that two observables located far from each other can be correlated. Indeed, the correlation between various pairwise commuting pairs of observables can be larger than anything that is possible in a classical theory, in which all observables commute with each other. This is shown by the fact that experiment, and quantum theory, violate the Bell inequalities. Correlation, even up to 100%, can occur between two classical variables (that is, between two commuting self-adjoint operators); the 'problem of Bell' arises when not all the observables commute with each other; the correlations concerned are, however, between pairwise commuting pairs of observables. Thus, suppose the system is in an entangled state. A measurement of anything at one point $\mathbf{x}$ in space does not alter the result that would be found at another point say $\mathbf{y}$, by measuring any observable at the same time. The result of the first measurement, by the first observer, cannot be known at $\mathbf{y}$ by the second observer. The first observer uses the reduced state (using the Copenhagen interpretation), whereas the second observer must use the original state: the state is not reduced for him, since the information available to the first observer is not available to him. In this way, using different states for observers with different information, the nonlocality of entangled states cannot be observed. This is why the word 'contextuality' is better than 'nonlocality'.

This question is distinct from those that arise in quantum information and measurement theory. I have nothing against this topical subject except the language in some papers, when they say that the effect achieved by the result of the experiment shows that quantum mechanics is nonlocal. Such a claim has been made, for example in the teletransportation of an electron state of unknown spin from A to B. This experiment requires that the space-time localisation of A should contain points in the backward light-cone of the space-time localisation of B; an essential ingredient is the transport of a classical bit of information at a speed less than that of light from A to B. Bell's inequalities do not arise in this situation.

With *EPR* [62] explained, other attempts to understand nonlocality by going beyond quantum mechanics are lost causes. Bohmian mechanics, if it is to agree with quantum mechanics (and experiment), needs to incorporate contextuality for its observables. For example, in [58] the concepts of momentum, angular momentum and even energy are not treated as observables; they are treated as quantum variables, whose statistical model is contextual. The positions of the particles are taken as the only observables in a theory of particles. However, to agree with quantum mechanics, the statistical model used to describe the position of a particle must be contextual, and so according to [58], the position variables are not observables either.

A Chapter discusses the Nelson theory and an earlier attempt by Wiener. I give brief accounts of these, trying to point out the errors. Later chapters cover quantum logic, trivalent logic, Jordan algebras and non-self-adjoint observables; these subjects, it is argued, are not worth pursuing (until you have tenure). In spite of my web pages on these topics, I still get e-mails pointing out apparent paradoxes arising in quantum mechanics. So I have expanded the pages to the present book, to include a few other lost causes covered by my site. I end with some more promising topics that seem possible to study.

Munchausen's Syndrome by Proxy

"... one of the most pernicious and ill-founded theories
to have gained currency in childcare and social services
over the past 10 to 15 years"
Lord Howe

Our first lost cause is not in physics, but in medicine; the pertinent point, however, is to do with elementary probability, and is similar to questions that arise in quantum mechanics. It thus provides a salutary lesson.

2.1 Munchausen's Syndrome by Proxy

would have continued to be regarded as a rare disorder in the domain of psychiatry, if its inventor, Sir Roy Meadow, had not attempted to justify its diagnosis by misusing statistics. This alerted scientists to the syndrome, and led to an unprecedented intervention by the Royal Statistical Society in a legal case. The theory itself was also reexamined, and on 15 July 2005, Meadow was struck off the medical register by the General Medical Council, but was reinstated on 17 Feb 2006 after a successful appeal. My conclusion is, and remains, that there is not and cannot be a scientific definition of the condition. It does not meet Popper's criterion of a scientific theory: it cannot be proved to be wrong by experiment. For, it is claimed by its proponents that no statement that a patient does not suffer from *MSP* can be taken as experimentally proved. There were also elementary mistakes in the diagnostics. The main error of omission in Meadow's analysis was the notion of conditional probability. It turns out that the quantum version of this idea is *quantum conditioning* which has also been misunderstood by some physicists; they are led to claim that there is no collapse of the wave-packet on measurement, contrary to the Copenhagen interpretation. From this, several schools of thought have been created, such as Bohmian mechanics, faster-than-light influences, quantum theory of the brain, and other lost causes. We use the case of Munchausen's to explain the use of conditional probability, as a prelude to Chap. (4). There we will explain the collapse of the wave packet.

2.2 Statistics in Human Affairs

When I was a student, the professor of statistics at Imperial College was G. A. Barnard. He warned us that EXTREME CARE must be used when applying statistical methods to human affairs. It is my contention that such care was not present in the case of Munchausen's syndrome; a misunderstanding of the notions of independence and correlation led Meadow to assign greater significance to the occurrence of multiple cot deaths than the statistics warranted. It is likely that this error was what induced him to invent the condition in the first place; it could have seemed to him to be the only explanation of otherwise remarkable data: the number of double cot deaths within the same family seemed to be far too high compared with the number of single cot deaths. This discrepancy is only there when the second cot death is independent of the first; in a model with correlation, there is nothing remarkable in the data. Meadow was not aware of the innocent possibility of correlation, and interpreted it as suspicious. Let us start with an account of Munchausen's syndrome by proxy *(MSP)*.

MSP is a psychological condition claimed to lead to infant- and child-abuse. It was so-named in 1977 by Sir Roy Meadow and is a variant of Munchausen's syndrome, invented and so-named by Sir Richard Asher in 1951. *MSP* has been diagnosed in many cases of cot death, and hundreds of mothers in the UK have been convicted of killing or harming their babies in the last 15 years. Thousands more, perhaps tens of thousands, have been deprived of their children on the basis of this idea. As a scientist, I am amazed that such a theory is accepted by doctors, since there is scant scientific evidence for it. There is, of course, evidence for *SIDS* (sudden infant death syndrome), infanticide and the deliberate harming of children. Indeed, David Southall has obtained secret video footage of some 30 mothers harming their children while they were in hospital [151]. It is the diagnosis of *MSP* to explain these events that must be regarded as not scientifically proved. One can ask, if they did not suffer from *MSP*, why did they harm their children? I shall suggest another possible explanation, which, while amazing, is scientifically possible.

The website of Dr. Marc Feldman [64] gives a definition of *MSP*. He first defines Munchausen's syndrome: "People with factitious [meaning false] disorders feign, exaggerate or actually self-induce illnesses. Their aim? To assume the status of 'patient' and thereby to win attention, nurturance and lenience that they feel unable to obtain in any other way." He goes on to define *MSP*: in the proxy form, someone else is harmed, and the benefits are vicarious. The doctor distinguishes between Munchausen's and malingering; in the latter condition, the patient creates real or false symptoms of disease *in order to obtain social benefits* such as disability payments. This is not the real purpose, we are told, of those suffering from Munchausen's. Similarly, in the proxy form, there is [supposed to be] a distinction between the psychological condition of those harming someone in their care in order to get attention and those doing it for money.

Some websites include in Munchausen's syndrome those harming themselves to get cash benefits, and in the proxy form, those harming others for the same purpose. I shall not use the terms in this sense.

It is the requirement of *intention* in these definitions that is difficult to test and impossible to prove scientifically. Intention is outside the realm of science, but has a place in legal questions. It seems to me that the part of the definition of the disease, involving intention to attract attention, was formulated to solve one of the mysteries of double cot deaths: in hundreds of such cases, the mother had no financial motive to kill her children, and since accidental death of the second child seemed [to Meadow] to be overwhelmingly unlikely, the death was murder and needed a motive. In the experiment of Dr. Southall, [151] about 39 mothers were persuaded to enter hospital with their children. Some were assigned private rooms, which, unknown to them, were equipped with secret video cameras. They were told that they were under observation (which was literally true). Dr. Southall found that 30 of the mothers were observed by video to inflict injury on their child, though other observers claim that the mothers' movements have an innocent explanation in seven cases. Even allowing that this were true, there remain 23 clear cases. This high "success" rate was explainable by the selection of the 39; the group consisted of mothers thought to be highly likely to harm the child. Perhaps all had previously been diagnosed, albeit informally, as suffering from *MSP*.

That all the mothers were suffering from *MSP* is not the only possible explanation, and is not insisted on by Dr. Southall. In the radio programme, "On the Ropes", broadcast on 9 August 2005, Dr. Southall was interviewed by John Humphrys, and he gave a neutral description of the events. However, one might take a more pro-mother view, even of the worse cases. It could be that a given mother had suffered from post-natal depression following the birth of a child; that this child had died a cot death; that after the second child was born, the mother again suffered from post-natal depression, which was compounded when she received a call from a famous pediatrician saying that her baby was in danger of sudden death; that they both should enter hospital for 'observation', with her staying over in the same room; that after some days, no treatment or even examination had been offered the baby; that she asked the pediatrician why, if the child had a life-threatening disease, the action of the hospital was so slow; that she was told that, while the Professor himself was sure that the baby's life was in imminent danger, the hospital staff were not convinced; that they wanted the room for their own patients, and unless some physical sign of distress became evident in the baby in the next few days, the baby would be sent and would then almost certainly die. In desperation, the mother decided to cause some injury to the baby, not to attract attention to herself, but to force the hospital to deal with the baby's life-threatening disease. While this scenario is incredible, it is just as likely as *MSP*.

While I was serving on the jury of a murder trial, the judge explained the difference between murder and manslaughter: to be murder, the accused must have had the *intention* to kill, or seriously harm, the victim. To convict, we the jury had to conclude beyond reasonable doubt that this intention was there. Similarly, to be guilty of stealing, an accused must have had the intention of permanently depriving the victim of the article taken. To be guilty of fare-dodging, the traveller must have "failed to buy a ticket in advance for the whole journey, with the intention of defrauding the Railway Company" [Bye-law, London Transport]. A new addition to this list is the crime of web-grooming: paedophiles [but, apparently, not others] who groom children on the web with the *intention* of meeting them and abusing them face 10 years in jail.

The difference between scientific proof and legal proof can be seen in a case of fare-dodging. A former student of physics at Imperial College decided that public transport should be free. He produced a pamphlet advising travellers to travel without any money and without a ticket, and to declare this at the ticket barrier. Provided that you give a name and address, you will be let through, he advised; London Transport will never check up and ask you for the money, as this action costs more than it brings in. Moreover, if you give your correct name and address, you will not be breaking the bye-law. After he had himself played this trick 150 times or so, he was arrested by the Railway Police and charged with defrauding the railway. At the magistrates' court, he chose to conduct his own case. Because any *intention* was internal to him, he asserted, the police could not prove that he had broken the bye-law by having the intention. They just did, replied the magistrate, SIX MONTHS IN PRISON!

It seems to me that in a cot-death murder trial, doctors who announce that the accused is suffering from *MSP* are usurping the functions of the jury. A similar view has been expressed by the Australian lawyer Michael Nott, in the Lawyers' Weekly of 26 September 2005. He remarks [128] that *MSP* is still used in Australia, but that this might soon change. He notes

In one significant decision for the Australian judiciary, Meadow's *MSP* was rejected by the Queensland Court of Appeal in the case *Rex v LM*, QCA 192[2004]. The Court ruled that *MSP* or factitious illness (*FII*) was inadmissible in evidence, and ruled out the testimony of doctors alluding to the alleged disorder. The court stated that *MSP/FII* was not a recognised medical condition, disorder or syndrome. In the case, Justice Holmes remarked that the Crown's use of *MSP* "explained nothing", "the Crown's argument was inherently circular and did nothing to prove criminal conduct". That legal reasoning was adopted by the UK High Court in the Family Division (A County Council v A Mother and A father and X,Y,Z children [2005] EWHC

31 Fam). In the case, Justice Ryder stated that he hoped *MSP* would be "consigned to the history books".

Michael Nott goes on

The Director of Public Prosecutions ACT, Richard Refshauge, said that the decision of the Queensland Court of Appeal on *Rex v LM* "... makes clear that if a woman is to be prosecuted for harming her children, it is not enough to put a label on it; facts are required to justify the case". "By labelling the woman in this way ... you are saying the woman is guilty, as the label creates the guilt. The problem is that labelling is not a process for determining guilt. [People] are convicted for the illegal acts that they do".

Yet the use of Meadow's beliefs continues in Australia, and in some states in the USA.

Some of the mothers accused of killing their own children because "they suffer from *MSP*" have formed their own website [120]. They are perplexed that once diagnosed to have the condition, cure is impossible. Feldman says that only one case of a cure is documented, though "many doctors consider [it] untreatable". Others say that a cure is difficult. The first step in all proposed cures is [93] the genuine admission to all the deceptions. This is very similar to the first step needed to cure witchcraft: admit it and denounce your accomplices. Huynh also says that the vast majority of *MSP*-sufferers deny that they have the syndrome. In slide 45, of [93] Huynh says, however, that "although many children return to the family and survive, there is no convincing case in the professional literature demonstrating successful treatment". In other words, it is [at present] not possible to find out whether or not the patient still has the syndrome. This admission leads one to the conclusion that *it was not possible to determine whether the patient had the "disease" in the first place*.

2.3 Meadow's Logical Error

About 300 women in the UK in the last two decades have been convicted of the murder or mistreatment of 1 or 2 of their infants, who died in the cot without known cause. A few years ago, about one baby in 8400 died a cot death in the UK every year, from well-off middle class families. Among poorer families with 2 or more children it is about 1 in 1,600. A serious misunderstanding of conditional probability by Sir Roy Meadow might already have led to widespread miscarriages of justice. He has said [118] that one cot death is a tragedy, two is suspicious and three is murder. He has calculated that the chance that two further cot deaths occur in a family of middle class in which one baby has already died a cot death is 1 in 73 million; this is so small (if it is chance) that it must be murder. Other medics have criticised his arithmetic, since it is evident that he has assumed that the second (and third) events are independent of the first, and simply multiplied 8,400 by itself. They are not independent; the rate of cot-deaths is higher in families in which there is a smoker, and among babies whose mothers put them to sleep face down. There is also a genetic component. So the rate of two further cot deaths is much larger than 1 in 73 million. Because of this error, and because some evidence of genetic susceptibility was witheld from the defence, four convicted mothers have been released by the Appeal Court, on the grounds that the murder convictions are unsafe. The Attorney General has started a legal review of all cases.

However, Meadow's error is much more serious than his rough assumption that the events, the two (or 3) cot deaths, are independent. The Royal Statistical Society criticised the incorrect use of statistics in one such case. They gave the following analogy. Suppose an unskilled archer fires one hundred arrows at a target of size 1 cm × 1 cm, drawn on a board 1 metre by 1 meter, and his aim is so bad that the arrows are distributed with uniform probability over the entire metre square. Assume (with Meadow) that each shot is independent of the others. There is then a chance of 1/10000 that a single arrow hit the target, but a reasonable chance, about 1/100, that one of the 100 arrows find the target. There is only a small chance that two arrows hit the target. If however, *after* the archer has fired his 100 arrows, one looks at the distribution, and chooses to draw the target round the two arrows that are closest together, one find a much better chance that both will lie inside a square 1 cm × 1 cm [this trick is known as "moving the goalposts"]. Meadow's mistake is not to recognise that in the case of cot deaths, one has moved the target by *selecting* a family with two cot deaths for investigation.

If the chance of a cot death is 1/1600, and given that there is some innocent cause giving rise to the failure of independence of the first and second death, then in a population as large as Britain's, one might expect several double cot deaths a year to arise *purely by chance*. One can then pick out a case of double cot death, but it would be wrong to say that the chance of an accidental death of the two babies in the case in hand is very low: the case is *selected by the very property whose chance one is assessing*. One should use conditional probability to assess the chances that *the selected sample* have various properties. The mathematics of this idea are explained in Chap. (3). For now, we do not need more than simple proportion.

Thus, among the population consisting of families with two cot deaths, one can ask what is the proportion in which there is medical evidence that one or both of them was murdered? Put this way, it is low, perhaps 3%, and almost certainly less than 10%, of the cases, given that there have been two deaths. The UK figures could be distorted by the spate of recent convictions. Ray Hill [90] says

"Obtaining reliable estimates bases on limited data is fraught with difficulty, but my calculations gave the following rough estimates. Single cot deaths outnumber single murders by about 17 to 1, double cot deaths outnumber double murders by about 9 to 1 and triple cot deaths outnumber triple murders by about 2 to one. So each successive death does give rise to some slightly increased suspicion, but to nothing like the extent that Meadow's law would imply. In particular, when multiple sudden infant deaths have occurred in a family, there is no initial reason to suppose that they are more likely to be homicide than natural."

It should be stressed that Hill's conclusion, that abuse is more likely if there are two deaths than if there is only one, is based on (limited) experimental data, and is not a theoretical necessity. Hill's calculations are given in [91]. Later on we shall see that even the contrary conclusion, that two deaths enhances the chance that they were natural, compared with only one death, is possible at least theoretically.

What is needed when a family suffers two SIDS? One can ask, has the family got a history of SIDS? One can also ask, what is the proportion in which one or both parents smoke? and so on, with each possible cause of death, given that there have been two deaths. But the fact that there have been two deaths, by itself, provides no evidence WHATEVER of murder, for the simple reason that the conditional probability that there are two deaths, given that there are two deaths, is unity, 1, 100%. One may then estimate the probability that it is accidental, in the absence of any other information, to be about 90%, assuming that the proportion with evidence of murder is about 10%. This is a far cry from 1 in 73 million. What the prosecution needs is evidence that a murder took place, not evidence that the chosen family suffered two deaths. It follows that if in a trial, the rarity of double deaths is assumed to provide supporting evidence, the argument is erroneous, and an appeal to the European Court should reverse the decision.

There have been many further cases in which there have been one or two cot deaths or other injuries, with no prosecution for murder or abuse, perhaps because the medical evidence is weak, but nevertheless the other children in the family have been seized by the social services. According to Margaret Hodge, then Commissioner for Children, the numbers of cases might run into tens of thousands. There have even been cases in which a prosecution for murder has failed, and the mother found NOT GUILTY, in which the remaining children have been taken into care for their safety. It looks as though the SS (Social Services) is reluctant to accept the verdict of a court of law. The SS say it happens because the burden of proof in civil cases is less than in criminal cases; the social services only need to show that on the balance of probability, the remaining children are in danger, not that this is true beyond reasonable doubt. However, we have seen that the chances of this are 9 to one against in the case of two deaths; the balance of probabilities favours the mother by 2 to 1 also if there have been three deaths. It is likely that an important factor is that the family courts act in secret, and the parents are not allowed to see all the evidence on which the care order is based [in a documented case, [20] the social worker destroyed one of three tapes made during disclosure sessions with a child; this seems to have been quite legal.] The parents and their advisors are not even allowed to discuss the case in any detail with their Member of Parliament or even the Commissioner for Children, without the permission of the Court. It is likely that the success of local MP Stuart Bell [20] in solving the Cleveland scandal has recently been countered by the *SS*, who make applications to the court to exclude everyone (except the *SS*) especially the MP. In a recent case, the state-appointed solicitor, purportedly acting for the child, as well as the solicitor for Kent Social Services, argued against any papers being sent to Margaret Hodge, and the judge concurred. That they were sent to the Solicitor General was pronounced to be contempt of court, even though this was then my good-hearted MP, Harriet Harman. The mother was not allowed to reveal the name of the doctor who had diagnosed her as suffering from *MSP*. I should say that some information was also released to the press. One of the objections put forward by social workers (according to the press) to speedy review of these cases is that, where adoption is pending, it will be interrupted if the case is found for the mother. Of course, the reader will understand that once a child is sent to an adoptive family (rather than just a foster home), the *SS* can argue that (in spite of a ghastly mistake having been made) it is now in the best interests of the child to remain in the new family. This argument has been made by Mrs. Hodge herself, who is a former social worker. Indeed, in a recent trial, a senior social worker pleaded guilty to perjury; she had said in a family court that the process of adoption of three children had progressed further than it actually had. She was given 12 months community service rather than 4 years in jail, because her statement was judged not to have changed the decision. She said that she did not know why she had told a lie. One might guess; it might have been to influence the outcome of the trial; or it might have been because she was worried that she had not followed the instructions of her line manager to get those kids adopted fast.

If it should happen that weak evidence of actual bodily harm is routinely enhanced by the claim that a double death is unlikely, then, as remarked above, the decision would be flawed: the 'balance of probability' may have been incorrectly assessed. I think that there should be a review of such cases as well as those in which the mother is in jail (55 cases or more). It is difficult to see how a review can be done (by the Solicitor General, assisted by the Commissioner for Children) if no papers on the case can be sent to them. In the case contested by Kent County Council, if the adoption process had been stopped, then Social Services Department might not have reached its target for adoptions that year, thus losing Government subsidy. Also, there is a great shortage of babies for adoption, and barren parents are tempted to treat as gods those wonderful people from Social Services who provide them with a child at last. These thoughts might well be entertained by the deprived birth parents [but not, of course, by this author]. Another thought they might have is that some remnant of the infamous 1:73 million might have been mentioned in the argument that, on balance, murder is more probable than not. One might ask whether the Social Services would be swayed at all by such a figure. The Appeal Court judge in a recent case asserted that the original jury was not swayed by the figure, simply because they had been warned by the judge who sat the case. And we all know that the Social Services of Kent and all other places consist of entirely professionally trained experts, who understand correlations and conditional probability, and can assess the balance of probabilities correctly, don't we? Even when not warned by a judge.

The BBC reported (May, 2004) that many of the Departments of Social Services, which have secured convictions for murder or abuse of babies suffering cot-deaths, do not intend to reopen any of the cases. No surprise there, then.

The National Society for the Protection of Children, the *NSPCC*, advocates that the rules of assessment should be altered, so that care orders can be issued when the carer's *MSP* is not just found to more probable than not, but its absence is not beyond reasonable doubt. This would, if implemented, reverse the burden of proof from the State to the parents. If this had been used up till now, the number of guilty parents not found guilty would be lower, and the number of miscarriages of justice would be much higher: recall that there is no known case of cure for *MSP*.

Following recent advice to put children to sleep on their backs or side, the number of cot-deaths in the USA, New Zealand, Tasmania and Scandinavia fell "dramatically" lately, and in the UK the number has been reduced by 70%. We might say, on these figures, that until recently, *about* 70% *of cot deaths are caused by placing the child to sleep on its face*. Other causes, such as genetic predisposition, premature birth, parental smoking, use of clamps in the birth ... must account for some of the remaining 30%. So the number of murders among double cot deaths could well be 10% at most, as found by Hill [91], during the period of many of the cases that worried Lord Howe [92]. So the balance of probabilities, with no other evidence, would then be 9 : 1 in favour of the mother; this figure is derived below. My source [2] cautions against automatically diagnosing murder in a case of one cot death, but does advise that if the cot death is discovered to be the second to be suffered by the family, then that might constitute evidence. This is a fallacious argument, as explained above; it is exactly the Prosecutor's Fallacy. This fallacy is the confusion between the two conditional probabilities

1. The chance that there are two cot deaths, given that the mother is innocent.
2. The chance that the mother is innocent, given that there have been two cot deaths.

The Royal Statistical Society point out that in assessing whether a rare event like two cot deaths is suspicious, we should not just find its probability, and if this is very low, come to the conclusion of murder. Instead, we should compare the chances of the very rare event, two murders, with the chance of an innocent reason for the deaths. This can be expressed in terms of frequencies, and is accessible to empirical evaluation. Suppose that a forensic scientist is asked to investigate the cause of a cot death in a family. With only this information, the probability that the death is murder can be estimated as $p(1) = m(1)/t(1)$, where $m(1)$ is the number of murders among a total of $t(1)$ families with at least one cot death. The revelation that there was an earlier death changes the chance of murder in the second case of death, in the absence of other evidence, from $p(1)$ to $p(2) = m(2)/t(2)$, where $m(2)$ is the number of murders of the later child among the total of $t(2)$ families with two cot deaths. $p(1)$ is an estimate of the conditional probability of one murder, given that there has been at least one death; $p(2)$ is an estimate of the conditional probability that the second death is murder, given that there have been two deaths. It is not clear that $p(2)$ is larger than $p(1)$. Thus is not clear whether the revelation that there was an earlier cot-death in the same family is evidence for, or evidence against, murder in the case at hand. Some people find that this result goes against intuition: after all, Shipman's murders were discovered because the statistics of the deaths of patients under his care were so unlikely. But thinking, not intuition, is needed, so as not to fall into the Prosecutor's Fallacy. When there have been two deaths, the earlier death could be evidence that a genetic defect is more likely than in the general population, thus reducing the chance that the second death is murder. Roughly, the argument hinges on whether the correlation between murders is higher, or lower, than the correlation between the deaths due to genetic (or other, innocent) factors.

A simple model shows how this can come about. Consider 400,000 mothers with two children; suppose that 200 mothers suffer from a genetic defect [= GD] leading to the death of a child within the first year, together with 25 different women who suffer from *MSP* [assuming there is such a disease]. We assume that deaths of any subsequent children of the GD mothers is 100% correlated with the first death, so is certain to happen. Suppose that a sufferer from *MSP* has a chance of 4/5 of killing the child within the first year, and that second, third … murders by such mothers are independent events with the same probability. Suppose that no deaths occur for any other reason. Suppose that a pathologist is asked to assess the probability that a particular cot death, of child A, is murder. There will be about 220 cases of one death, this being all the babies of mothers with the genetic defect, and about 20 of the 25 MSP mothers. On this evidence alone, the chance that child A was murdered is $20/220 = 1/11$. If the pathologist digs deeper, and finds out that A's older sibling also died a cot death, then the information shows that child A is a member of a population of 216, this being the 200 GD mothers who lost both children, and 16 out of the 20 *MSP* mothers who had killed the first baby. So the probability that A was murdered is now 16/216 [with the extra information], which is even less than 20/220.

This example shows the following features.

1. The total population of 400,000 mothers does not appear in the conditional probability, and the fact that a cot-death occurs rarely (one case per two thousand) is not relevant.

2. The number of double deaths is nearly as large as the number of single deaths, about one in two thousand and not one in four million as it would be if the events were independent. Evidence like this can be used to show that the 2 deaths are not independent, but the reason for the dependence is not that all 200 GD-mothers suffer from a compulsive murder syndrome, invented just to explain the high rate of double-deaths.

3. The probability that a woman with one cot death has murdered the child is $1/11$, whereas if it is revealed that the family also suffered an earlier cot death, then the chance is reduced to $2/27$. Thus, the suggestion in [2] that the revelation that the family has already suffered a previous cot-death provides evidence of murder may not be justified, even though it is official advice of a reputable body, the American Academy of Pediatrics .

Bear in mind that the probabilities estimated here are in the absence of any clinical evidence of murder. The calculation also postulates that MSP exists, at the rate of 25 in 400,000, but that it is not known which of the mothers is "MSP-positive" and which suffer from GD.

There is a danger of committing the Prosecutor's Fallacy in assessing the significance of DNA matching and finger-print evidence. For example, suppose that a sample of a finger-print is obtained at the scene of a crime, and the data base of 10 million samples is searched for the closest match, which is found to be Mr. B. Then this by itself is not much evidence against Mr. B, unless it is an exact match. This is rarely the case; these days, the match is sought by computer using fuzzy logic, and it reveals the closest match even when the culprit's data are not on the data-base. A close match can be used as a tool for further investigation, and also in conjunction with other evidence that Mr. B is involved in the crime; but to argue that the chance of such a close match in one in 10 million is to fall into the same error as Meadow. However, a difference between the sample and all of Mr. B's prints or DNA can correctly prove the innocence of Mr. B.

It is a common occurrence that apparently startling correlations or other statistical patterns lead observers to erroneous conclusions; often, apparently remarkable coincidences are quite likely to have occurred by chance. Such a case is the occurrence of ley lines. This saga concerns the location of quasi-religious Celtic monuments in pre-Christian England and France. If the locations are plotted by dots on a map of Europe, then looking at it by eye, one notices (if told to look for it) that very often, three, or even four of the sites lie in a straight line. The line joining them might even cross the Channel. So the theory arose that the ancients were able to conduct geodesy on a large scale. However, an analysis by D. G. Kendall [100, 149] shows that the accuracy of the alignment was not remarkable, compared with the likely alignments of a Poisson random field of points. So the set of data is consistent with a random selection of each site. This nice piece of work did not, however, change the minds of the many followers of the cult of ley lines.

A recent note to social workers has suggested that *MSP* is a bad name, and that other terms should be used. For example, recently the terms "fabricated or induced illness" have been used. We can expect, then, that the existence of *MSP* as a cause of double cot-deaths will decline, but that under a new name, it will continue to be advocated by its inventor and other adherents.

3

Elementary Probability

> The Truth is –*Education is no good unless you know it.*
>
> W. C. Sellar and R. J. Yeatman, **And Now All This**
>
> Methuen & Co., London, 1932.

3.1 Basics

There is no other simple mathematical theory that is so badly taught to physicists as probability. We hope to put this right in this chapter by discussing the easiest case, when the sample space is finite.

Two things annoy probabilists about the way physicists are taught probability. One is the confusion between the sample space and the values taken by some random variable; and the other is the use of frequency to define probability. To take the first point, consider a die with sides labelled 1 to 6. The sample space is then $\Omega := \{1, 2, 3, 4, 5, 6\}$, and the points of Ω are called *outcomes*. This does not mean that when a player rolls a five he will win 5 units. The amount he wins might be agreed between the players, to be some specified function on this set, i.e. a map $X : \Omega \to \mathbf{R}$. Any such function is called a random variable. For example, in a horse race, the sample space is the field (of horses) and the "outcome" is the winner; the "odds" offered by a bookie for each horse in a race do not sum to 1, and so cannot be the probability that it will win. Rather, the odds defines a random variable whose value is the winnings per unit investment. For example, odds of 5 to 2 on a horse means that if £ 1 is laid, and the horse wins, then £ 3.5 is returned to the backer.

Physicists often use the values taken by a chosen random variable X as labels of the outcomes. This is alright, unless X takes the same value at two different points; then we lose the distinction between some different outcomes.

Moreover, when they want to discuss several random variables, such as the winnings of various players when the die is rolled, they get confused about things. We therefore insist that there is a sample space, Ω, a typical point of which is denoted ω. Some random variables, that is, functions of ω, representing observables which are of interest, should also be specified. That a random variable, defined as a function of ω, is in fact random, is no mystery: the value of X which is observed after a sample ω is supplied, is $X(\omega)$, and this is random because ω is.

The second question, what is meant by probability, was vague in even the mathematics literature until the work of Borel and Lebesgue; the definitive notion was given by Kolmogorov [108]. A model probability space is (in the finite case) given by a finite sample space Ω and a probability on it; that is, a function p of ω, satisfying the axioms

1. $p(\omega) \geq 0$; $p(\omega)$ is termed the probability that ω occurs.
2. $\sum_{\omega \in \Omega} p(\omega) = 1$.

Note that we do not say that each outcome is equally likely; moreover, no experiment is mentioned in the definition, unlike in the definition using frequencies. The contact with experiment is made in the same way as in all applied mathematics: we adopt a model, $\{\Omega, p\}$, called the hypothesis, and then test it against an experiment. We accept or reject the model, depending on whether its predictions are borne out by the data. Indeed, probability has a very refined theory of when to accept a model or not, this being the theory of statistical hypothesis testing. If a model is rejected, suggest another model, using the data as a guide. This may differ from the first in the choice of p, and also in the choice of Ω. We can perform a test of the model, and apply estimation theory, when we have a large supply of samples of the system, which are produced in the same way. If we only have one sample, we might be able to set up a model $\{\Omega, p\}$ based on a theory or experience. Then $p(\omega)$ would be a rational measure of our confidence that the sample to hand is ω. We shall not develop the idea of rational belief here, but shall limit our discussion to the case when a stream of samples is available. This will cover both the classical and quantum applications we have in mind.

As an example of a probability on the space $\{1, 2, 3, 4, 5, 6\}$, suppose that $p(1) = p(2) = \ldots = p(6) = 1/6$: all outcomes are equally likely. Then we say that p is uniform, or that we have a "fair" die. As remarked, this is far from the general situation. An even simpler case is that of a fair coin, which when tossed shows head $= 1$ or tail $= 0$ with equal probability. This is modelled by the sample space $\Omega = \{0, 1\}$ with probability $p(0) = 1/2$, $p(1) = 1/2$. This p is called the Bernouilli distribution.

Returning to the general case, if $\omega \in \Omega$ is such that $p(\omega) = 0$ we say that ω is impossible, while if $p(\omega) = 1$, we say that ω is sure. Clearly, this depends on p. A random variable X is said to be sure if it has the same value whatever outcome turns up, that is, if $X(\omega)$ is independent of ω. Thus, for a given probability model, say (Ω, p), the sure functions are one-to-one with the set **R** of real numbers.

An *event* E is a subset of Ω; an outcome, say ω, can be identified with a special type of event, consisting of the one-point set $E_\omega = \{\omega\}$. Now let $E \subseteq \Omega$ be any event. Given a sample ω, such as from a beam, we say that the event E has occurred, or happened, if our outcome ω lies in E. For example, if we roll a die, then the subset $\{2, 4, 6\}$ is the event "the outcome was even". We may extend the definition of p from points to subsets, i.e. from outcomes to events, by

$$p(E) := \sum_{\omega \in E} p(\omega). \tag{3.1}$$

Events commonly arise from random variables X as their *level sets*. Thus, if X takes values $x_1, \ldots, x_n$, then the set $E(x_1) = \{\omega \in \Omega : X(\omega) = x_1\}$ is the event that can be described by saying "X took the value x_1". We can repeat this definition for each possible value that X can take, to get sets where X takes different values. This divides Ω into disjoint events, that is, non-overlapping subsets:

$$\Omega = E(x_1) \cup E(x_2) \cup \ldots E(x_n) \tag{3.2}$$

where n is the number of different values taken by X. A division of Ω into non-overlapping subsets is called a *partition of* Ω. If we measure X, we know which part of the partition that outcome lies in. Let us write $E(x_1) = E_1$ etc.

Two events E and F are said to be *independent* if $p(E \cap F) = p(E)p(F)$. Note that this involves p as well as the events themselves. This is not the same notion as being mutually exclusive; the latter means that E and F are disjoint sets: $E \cap F = \emptyset$, the empty set. Two random variables X, Y are said to be independent relative to p if each level set E_i of X is independent of each level set, F_k say, of Y.

If each part, E_i, $i = 1, \ldots, n$ of the partition of Ω given by the level sets of a random variable X consists only of one point, then we can find out what ω occurred just by measuring X. Such a random variable is said to *separate the points* of Ω. If X does not separate the points, then we only get limited information about ω by measuring X. We should still describe our knowledge by a probability, but one that is *conditioned* by our new knowledge, that $X(\omega) = x_1$, say. So we know that ω lies in E_1, and the probability that it lies outside this set is now zero. It would be wrong to assume that all the points in E_1 are then equally likely; indeed, the new probability of ω is proportional to the original probability, $p(\omega)$, provided $\omega \in E_1$; otherwise it is zero. Thus we get the formula:

$$p(\omega | X(\omega) = x_1) := \frac{p(\omega)}{p(E_1)}. \tag{3.3}$$

We shall call this Bayes's formula, though this name is often used to describe the simple consequence, namely, the estimation

$$p(E_1) = \frac{p(\omega)}{p\left(\omega|X(\omega) = x_1\right)}. \tag{3.4}$$

The left-hand side given by Eq. (3.3) can easily be shown to obey the axioms of probability; it is called the *conditional probability given that X takes the value x_1*. We see that Bayes's formula spreads the total probability, 1, over the set E_1, the given information, rather than over the whole of Ω. It is clear that this increases the probability of outcomes in E_1 compared to the original p, and that this increase is very large when $p(E_1)$ is very small. The subjective nature of probability stems from the fact that it is changed (by Bayes's rule) when information becomes available. Bayes's rule is fully supported by experiments. Failure to appreciate the difference between p and its conditioned form Eq. (3.3) led to the grave error in assessing the probability of murder in cot deaths, as we saw in Chap. (2).

There is a quantum version of Bayes's formula, related to the postulate known as 'collapse of the wave function'. Deniers of this postulate, a key element of the Copenhagen interpretation, are also in error.

3.2 Probability Distributions

An important concept is that of probability distribution. The definition is given in terms of the underlying p and does not depend on experiment. Suppose that (Ω, p) is a probability space, and X is a random variable on Ω, taking the values $x_1, x_2, \ldots, x_n$. The probability distribution of X is the rule that assigns the probability $p(E_i)$ to the value x_i, $i = 1, 2, \ldots, n$. This we shall denote by $p_X(x)$, thus:

$$p_X(x) := p\{\omega : X(\omega) = x\}. \tag{3.5}$$

It is a probability on the space of level-sets of X. The probability distribution should be distinguished from the frequency distribution, which comes next.

Most physicists and some statisticians begin the study of probability with *frequency* distributions. We have a "variate", X, something accessible to experimental measurement, that varies randomly, and we measure the number of times in a long run of r "independent" trials, that X takes its possible values $\{x_1, \ldots, x_n\}$. If n_i is the number of times x_i is found then the frequency of x_i in that run of trials is defined as $f_i := n_i/r$. The collection of such fractions is called the frequency distribution of X. It is the experimental approximation to the probability distribution of X; as we saw above, the probability distribution can be calculated relative to our chosen probability model. The model, known as the "hypothesis H", is rejected, if the discrepancy between the experimental values of the frequencies, and the theoretically correct values, $p_X(x)$ is *significant*. There is a well-developed theory of *significance testing*. In its simplest form, the hypothesis H involves the choice of Ω and a probability p on Ω. We test the hypothesis by measuring a random variable X for a sample ω, getting a value x say. This can be compared with the "mean", also called the "expectation", predicted by the hypothesis H. As usual, we define the *mean* of a random variable X relative to p to be

$$\mathbf{E}_p[X] := \sum_{\omega \in \Omega} p(\omega)X(\omega). \tag{3.6}$$

The mean gives us some idea about the location of X. The variance,

$$\mathcal{V}(X) := \mathbf{E}_p[(X - \mathbf{E}_p[X])^2], \tag{3.7}$$

on the other hand, gives us some idea of the *spread* of X around the mean. If the observed value x is so far from the mean that the chances, under H, of getting it, or any worse value, is less than say $s = 1\%$, we say that the measurement was significant at the 1% level, and reject H. Otherwise, we stick with our model for the time being. This can give rise to a *false positive*, in which a quite wrong choice of p is not detected. This possibility is made less likely if we raise the level of significance, s, to, say, 5%. On the other hand, raising to this level means that some data which are not significant at the 1% level become significant; if H be true, one of twenty experiments is likely to result in the 5% tail of the distribution. This type of error, the *false negative*, is reduced by combining the information got by a run of tests.

In the calculation by Meadow, Chap. (2), who arrived at the notorious figure that cot-deaths of two babies occurs at the rate of 1 in 73 million, there are some of the ingredients of a standard statistical test; but he goes badly wrong. His original hypothesis H was that the chance of one cot-death was 1 in 8400, and that subsequent deaths are independent of the first. When this was criticised as too small, he was happy to replace the figure of 1 in 8,400 by 1 in 1000, as an apparently generous gesture to the *MSP* mothers. This led of course to the smaller figure of 1 in a million for two deaths, under H. This figure was accepted by the judges in one of the cases of appeal; the judgement said that 1 in 73 million was not intended as a precise figure, but was in the same ball-park as the correct figure, between that and 1 in a million. Naturally, if 1 in a million is interpreted as having the status of a significance level, then *any* double cot-death is very significant; even this figure is so much less than the usual significance levels, 1% or 5%, that one should immediately reject H. However, the figure is the probability of one event, given H, not the conditional probability of murder, given two deaths. So the line of argument is not related to a significance test for murder. One can apply the standard significance test to H if we use the *full* data on the numbers of single and double cot-deaths; it turns out that when this is done, one must reject H. Since the probability of a single cot-death is well estimated to be $1/1,600$ (or so), we should keep this part of H and reject independence. This tells us is that there is a correlation between the first and second death; it does not imply in any way that either death was murder.

Generally, if a hypothesis H is rejected in a significance test, we use estimation theory to modify the model. There is a full and sophisticated theory of estimation, which we omit here.

The meaning of "independent trials" in a statistical experiment is not mathematical, but expresses the idea that the experiment should be conducted so that there is no way the results of the first i trials can effect the outcome of the $(i + 1)^{st}$. We shall say that such variates are *physically independent*. There is a natural way to construct a model of a sequence of r physically independent trials; we start with a probability model (Ω, p) for one trial: we take the sample space for the sequence to be the set of ordered $r-$tuples of outcomes, namely

$$\Omega := \Omega \times \Omega \times \cdots \Omega \ (r \text{ factors}). \tag{3.8}$$

Thus a sample point is now $\boldsymbol{\omega} = (\omega_1, \dots, \omega_r)$, with each $\omega_i \in \Omega$. We furnish the product space with the product probability

$$P(\omega_1, \omega_2, \dots, \omega_r) := p(\omega_1)p(\omega_2) \dots p(\omega_r). \tag{3.9}$$

The model consists of r exact copies of Ω of our original model (Ω, p) embedded in it, in the sense that we can define r random variables X_i on $\boldsymbol{\Omega}$, each a copy of X, thus:

$$X_i(\boldsymbol{\omega}) := X_i(\omega_1, \omega_2, \dots, \omega_r) := X(\omega_i).$$

Then one shows easily that all the X_i, $i = 1, \dots, r$ are *statistically* independent random variables on the product space $\boldsymbol{\Omega}$, and that they all have the same distribution as X. These thus form a good model for a run of r trials of physically independent experiments.

It is possible to some extent to test whether the trials of a sequence are independent; in the construction just given, we included independence in the hypothesis, in the product form taken for the measure in Eq. (3.9). If the probability of the result, or a worse result, is greater than $s\,\%$, we reject the hypothesis H, which includes the assumption that the trials are independent. Even if the trials are not independent, the sample space of the sequence would still be correctly given by (3.8), but we would need to choose a different probability on it.

It is always better to test a hypothesis by a run of r independent trials, rather than testing just one sample. Typically, we form the "average" of the values, namely

$$\overline{X} = r^{-1} \sum_{i=1}^{r} X_i. \tag{3.10}$$

This is a random variable on the product space, whose mean is the same as that for a single sample, under H. We then test whether the value obtained is close enough to the mean. It can be shown that the larger r is, the smaller is the chance of a false positive (for the same value of s). This arises because $\overline{X}$ has the variance $r^{-1/2}\mathcal{V}(X)$, which becomes very small as r becomes large.

Thus a large value of $|\overline{X} - \mathbf{E}[\overline{X}]|$ becomes very unlikely. The possibility of constructing models of sequences of a system from the model of the system itself is essential in any branch of science; we must be able to analyse the data into *reproducible results*. In the case here, this is expressed by saying that random variables form a *tensor category*.

We claimed earlier that probability is no different from any other subject in applied mathematics. Indeed, we first propose a model, based on theoretical insight, or trials; we then do experiments; if the predictions of the model are significantly different from the experiment, we reject the model, and then modify it in the light of the results, and try again.

An important class of random variables are those that take the value 0 or 1, the so-called *questions*. Let Q be a question; it defines an event $E(Q)$ defined as the subset of points ω such that $Q(\omega) = 1$. Thus, Q is the *indicator function* of the event $E(Q)$. The answer to the question, "did $E(Q)$ happen?" is yes if the outcome ω is such that $Q(\omega) = 1$ and no if $Q(\omega) = 0$. A question Q obeys the equation $Q^2 = Q$ (equality meaning both sides are equal as functions of ω). A space Ω_0 with only two points, 1 and 0, with indicator functions $Q(1)$ and $Q(0)$, is very suitable for modelling the toss of a coin. The probability then has the form $qQ(1) + (1 - q)Q(0)$, for some $q \in [0, 1]$. We take q to be the probability of getting "yes". If we take $q = 1/2$, we have the Bernouilli model.

The example of a stream of samples gives rise to a *stochastic process* with discrete time. In its simplest form, a stochastic process is a family of random variables, $\{X(t)\}$ labelled by time $t \geq 0$. Consider repeated, independent tosses of a coin, for a finite number r of tosses. The general construction of the product space and product probability, given above in Eq. (3.8) and (3.9), then says that an outcome is represented by terminating binary fraction $\cdot\omega_1\omega_2\omega_3 \ldots \omega_r$, where each binary digit, equal to 0 or 1, is the outcome of the corresponding toss. The sample space is the product-set $\Omega = \Omega_1 \times \Omega_2 \ldots \Omega_r$ and independence is expressed by the product law:

$$p(\cdot\omega_1\omega_2 \ldots \omega_r) = p(\omega_1)p(\omega_2) \ldots p(\omega_r). \tag{3.11}$$

Suppose that the probability of getting "heads", denoted 1 in the above binary, in one toss, is q. One form of the *law of large numbers* states that experimental frequency of getting a head in these r independent trials, namely the number of all the ones that have occurred, divided by r, converges to q in the probabilistic sense, as r becomes large. This deep result is sometimes used by frequentists to justify the use of long runs of experiments to *define* probability. This idea is circular, however; the "convergence" is not that of analysis. Convergence in the probabilistic sense means that the probability that the observed frequency be far from the expected q is very small; the probability in question itself is derived from Eq. (3.11), which involves the number q we are attempting to define. This is why there is not a single theorem that can be proved using the frequentist definition of probability.

In any set Ω, a useful concept is the *indicator function* $\chi(E)$ of a subset $E \subseteq \Omega$; this is a function of ω, equal to 1 if ω lies in E, and zero if ω lies outside E. When Ω is finite, with N points, then the indicator function Q_j of the points $\omega_j, j = 1, \dots N$ form a linear basis for all random variables: any X can be written as a linear sum of them, the coefficients being the observed values of X:

$$X(\omega) = \sum_{j=1}^{N} X(\omega_j) Q_j(\omega). \tag{3.12}$$

It follows that the set of random variables is a vector space of dimension N. It is also an algebra, in that the product of two functions of ω, pointwise, is also a function of ω, and so is a random variable. This idea can be generalised into the concept of an *information algebra*. Suppose that we are able to measure the random variable X, which does not separate the points of Ω. So there are outcomes, ω_1, ω_2, which yield the same measurement for X. The level sets of X divide Ω up into equivalence classes; two points in the same level set are equivalent. We have thus a *coarse graining* of Ω, a partition, and this implies that there is a lack of full information associated with X. We cannot improve the information by measuring a function of X, such as X^2 or X^3. The set of all functions of X is a vector space, and a basis is given by the indicator functions of the level sets of X. By taking unions of these level sets, and adding the empty set, we arrive at a Boolean ring of sets: the whole space Ω is a member; the union of two members is a member, as is the intersection; the complement of any member is also a member. This Boolean ring is said to be generated by X, and will be denoted $\mathcal{B}(X)$. The use of the word "ring" arose because there is a natural way to define a product ($=$ the intersection, $E \cap F$) and a sum ($=$ the symmetric difference $E \cup F - E \cap F$) of two sets E and F; these obey the axioms of a ring. This ring structure plays no role in the sequel, and neither does the structure of Boolean algebra; this is defined from the ring structure by adding scalar multiplication by the two-element field $(0, 1)$. We shall use the term *information channel* for $\mathcal{B}(X)$. The channel is said to be *noisy* if X does not separate the points. The functions $f(X)$ of X form an algebra (a vector space with a multiplication), which we denote by $\mathcal{A}(X)$; we reserve the term *information algebra* for this. Its elements f are seen to be measurable functions relative to $\mathcal{B}(X)$; that is, the inverse image $f^{-1}(a, b)$ of any real interval (a, b) lies in $\mathcal{B}(X)$; in finite spaces, a function Y is measurable (relative to $\mathcal{B}(X)$) if it is a linear combination of the indicator functions of the level sets of X. Such a random variable can be measured by measuring X (since it is, in fact, a function of X). Now the idea, measurability, was originally introduced by Lebesgue to remove pathology from integration theory over the real line; it is here revealed to be a fruitful heuristic in information theory even in finite spaces.

3.3 Moments

We have already defined and used the mean and variance of a random variable. The mean is also called the first moment. The variance of X, which measures the size of the spread of the distribution around the mean, defined as in Eq. (3.7), can also be written

$$\mathcal{V}_p[X] := \mathbf{E}_p[X^2] - (\mathbf{E}_p[X])^2. \tag{3.13}$$

This can thus be expressed in terms of the moments

$$m_j[X] := \mathbf{E}_p[X^j], \qquad j = 1, 2, \ldots. \tag{3.14}$$

Thus

$$\mathcal{V}_p[X] = m_2 - m_1^2 = \mathbf{E}_p\left[(X - m_1)^2\right]. \tag{3.15}$$

Another important idea is the determination of the distribution from its moments. Naturally, in (3.14), X^j is a random variable, whose expectation is given by (3.6) with X replaced by X^j. When Ω is finite, as here, it can be shown that the distribution of X, $p_X(x)$, is determined by the sequence of its moments. This is expected, since there are infinitely many moments, and only finitely many unknowns, $p(\omega)$. Even so, we have to show that there is enough information in the moments to give a unique result. Indeed, if $|\Omega| = \infty$, there are moment sequences that do not determine the measure on Ω uniquely. An easier result is that the moments are determined by the distribution:

$$m_j(X) := \mathbf{E}[X^j] := \sum_{\omega \in \Omega} p(\omega) X^j(\omega)$$

$$= \sum_{i=1}^{n} \left(\sum_{\omega \in E_i} p(\omega) X^j(\omega) \right)$$

$$= \sum_{i=1}^{n} x_i^j \left(\sum_{\omega \in E_i} p(\omega) \right)$$

since X takes the value x_i if ω lies in E_i

$$= \sum_{i=1}^{n} x_i^j p_X(x_i).$$

It should be said that not every sequence of numbers can be the moment sequence of a distribution. We see from Eq. (3.15) that the variance of any random variable is the mean of a square, and so is non-negative. Thus we see that the inequality

$$m_2(X) \geq (m_1(X))^2$$

must hold for any random variable X. There are infinitely many other inequalities that must hold for a sequence to be a moment sequence. We shall see that a similar set of inequalities holds also in the quantum case. The underlying reason for these inequalities is positivity: the expectation of any non-negative random variable is non-negative; thus, for all complex $\{\alpha_i\}$, $i = 1, \ldots, \alpha_n$, we get

$$\mathbf{E}_p\left[\left|\alpha_i X^i\right|^2\right] \geq 0;$$

This leads to the condition on the moments $m_j(X)$:

$$\sum_{i,j} \overline{\alpha}_i \alpha_j \mathbf{E}_p[X^i X^j] = \sum_{i,j} \overline{\alpha}_i \alpha_j m_{i+j}(X) \geq 0, \qquad (3.16)$$

for all complex numbers $\{\alpha_i\}$, $i = 1, 2, \ldots, N$. Note that in Eq. (3.16) $\overline{\alpha}$ is the complex conjugate of the complex number α, not an average as in Eq. (3.10). One can prove that this condition, for all N, is sufficient as well as necessary, for the sequence $\{m_j\}_{j=1,2,\ldots}$ to be the sequence of moments of a random variable.

Now consider the case when we have two random variables, X and Y, taking respectively the values $x_i, y_j; i = 1, 2, \ldots n, j = 1, 2, \ldots m$, say. Each divides Ω into level sets, and we can consider the double level sets

$$E(x_i, y_j) := \{\omega \in \Omega : X(\omega) = x_i \text{ and } Y(\omega) = y_j\}. \qquad (3.17)$$

We can define the joint probability distribution, the probability that X takes the value x_i and Y takes the value y_j. We shall denote this distribution by $p_{X,Y}(x_i, y_j)$. Again, the multiple moments $\mathbf{E}_p[X^i Y^j]$, $i = 1, 2, \ldots, j = 1, 2, \ldots$ can be defined in terms of the joint distribution, and the full double sequence of joint moments determines the joint distributions. There are now an infinite series of inequalities that must hold between the various joint moments; these follow from the requirement of positivity; namely, the mean of any square is positive:

$$\mathbf{E}_p\left[\left|\sum_{i,j} \alpha_{ij} X^i Y^j\right|^2\right] \geq 0 \qquad \text{for all complex values of } \alpha_{ij}.$$

This includes the conditions on each sequence, the moments of X, and the moments of Y; but some constraints also hold on the mixed moments. For example, writing $\mathbf{E}$ for $\mathbf{E}_p$ for simplicity, we get Schwarz's inequality:

$$(\mathbf{E}[XY] - \mathbf{E}[X]\mathbf{E}[Y])^2 \leq \mathcal{V}[X]\mathcal{V}[Y] \qquad (3.18)$$

follows. This has the important consequence that the covariance matrix

$$\mathcal{C}(X, Y) := \begin{pmatrix} \mathcal{V}[X] & \mathbf{E}[XY] - \mathbf{E}[X]\mathbf{E}[Y] \\ \mathbf{E}[XY] - \mathbf{E}[X]\mathbf{E}[Y] & \mathcal{V}[Y] \end{pmatrix} \qquad (3.19)$$

is positive semi-definite. Equivalent to this is to say that the *correlation coefficient* of X and Y, defined as

$$c(X, Y) := \frac{\mathbf{E}[XY] - \mathbf{E}[X]\mathbf{E}[Y]}{(\mathcal{V}[X]\mathcal{V}[Y])^{1/2}}$$

lies in $[-1, 1]$. If it is negative, we say that X, Y are anti-correlated, and if it is positive, we say that they are correlated. It is zero if X and Y are independent, which is an example of being uncorrelated. If two random variables X, Y are (positively) correlated, then positive values of X often but not necessarily always occur at an outcome ω where Y is also positive; and negative values also tend to occur together. In statistics, we say that there is an "association" between the variates X and Y. J. S. Bell discovered a new inequality that must hold between certain combinations of correlations among four random variables. This is the Bell inequality, and it has the remarkable property of not being true in quantum mechanics. Since experiments carried out by Aspect, Dalibrand and Roger *et al.* [10] with photons in a suitable state was found to violate the Bell inequality, we conclude that quantum probability cannot be described by ANY classical probability model. This has been regarded as a paradox, which has led to much heart searching and wrong analyses. To waste time on research into this is a lost cause, as its explanation is quite elementary, and is given in Chaps. (6,8).

3.4 Probability with Two Observers

This topic is usually omitted from the books, even those that go way beyond this essay in technical difficulty. It shows firmly that probability has a subjective part, in that it depends on the information available to the observer, as well as the probability p underlying the model. So two observers with different information will, quite rationally, assign different probabilities to the same event. Suppose that (Ω, p) is a probability space, and that there are two observers, A=Alice and B=Bob, who know p. Suppose that Alice can measure the random variable X, and that B can measure the random variable Y. In general, the level sets of X and Y will be different. After making a measurement, they each will use Bayes's formula (3.3) to condition (Ω, p) by the information received. Their calculation of the probability that the outcome is the point ω will be different; even the sample spaces will not be the same. For Alice, who found that $X(\omega) = x_1$, the sample space will be the level set $E_1 := \{\omega : X(\omega) = x_1\}$ while for Bob, who found that $Y(\omega) = y_1$, the sample space is now the level set $F_1 : \{\omega : Y(\omega) = y_1\}$, a different set in general. More, the conditional probabilities they assign to a possible common element are not the same, unless $E_1 = F_1$. Yet Bayes's rule does predict the correct experimental frequencies of guessing the right answer for two observers playing dice, cards, etc.

Since A can measure X, her potential information is summarised by the family of conditional probabilities $p(\omega|E)$ on Ω as E runs over the level sets of X. But B, who can measure Y, has access to the different family of conditional probabilities $p(\omega|F)$ as F runs over the level sets of Y. This structure appears in game theory; there the basic sets of a coarse graining are called information sets, and are different for each player. The underlying sample space Ω is still there, the same for both players; they differ in their ability to extract information. A must use the Boolean ring $\mathcal{B}(X)$ and B must use the ring $\mathcal{B}(Y)$; the original probability p on Ω gets conditioned by a measurement, and this leads to different knowledge about the sample being obtained by the observers. Equivalently, A can measure random variables in the algebra $\mathcal{A}(X)$ and B can measure the random variables in $\mathcal{A}(Y)$. Recall that $\mathcal{A}(X)$ is the algebra of functions of X. This idea of specifying the algebras, rather than the information sets, can be generalised to quantum theory.

Game theory proper involves not only (Ω, p) and the two information sets. The purpose of the game is for each player to develop a strategy that maximises the expected value of his reward function. This feature does not arise here; but the introduction of an information set, or equivalently, an algebra of measurable functions for each observer, is still necessary.

Naturally, if there are more than two observers, we introduce a different Boolean ring for each observer, representing the information accessible to that observer.

We shall see that the same partial subjectivity arises in quantum probability when there are two or more observers. The most complete analysis of this idea arises in *local quantum theory* where an algebra of observables is assigned to each bounded open set of space-time; this algebra contains the observables that can be measured by apparatus lying in that region of space, and active over any interval of time inside the region.

The failure to recognise the subjective nature of quantum probability has led to confusion over the collapse postulate of quantum mechanics.

3.5 Statistical Models

Statisticians working in the human sciences prefer to work with the distributions of some accessible random variables, rather than to define a sample space and a probability on it. For example, they measure the height X and weight Y of a random sample of people, and find the frequency distributions. In practice, the values of X and Y are coarse-grained, as they are measured only to the nearest cm. or gm. say, and the samples are assigned to intervals in these units, called *bins*. The experimenters also need to measure the joint frequency distribution of X and Y, from which all moments and cross-moments can be found. The joint frequency distribution must be related to those of X and Y, since the distributions of X and Y are the *marginals* of the joint distribution:

$$f_X(x_i) = \sum_j f_{X,Y}(x_i, y_j) \qquad f_Y(y_j) = \sum_i f_{X,Y}(x_i, y_j) \qquad (3.20)$$

We also need the usual condition

$$\sum_{i,j} f_{X,Y}(x_i, y_j) = 1. \qquad (3.21)$$

These are the simplest examples of the *Kolmogorov consistency conditions*. It is easy to prove these conditions, if there is a probability space (Ω, p) and a pair of random variables $X(\omega)$ and $Y(\omega)$ whose distributions, and joint distribution, are the given ones. The converse also holds: given distributions f_1, f_2 and $f_{1,2}$ obeying the consistency conditions (3.20) and (3.21), with $1, 2$ replacing X, Y, then there exists a probability space (Ω, p) and two random variables X, Y on it, taking the values x_i and y_j such that

$$f_1(x_i) = p_X(x_i), \quad f_2(y_j) = p_Y(y_j) \text{ and } f_{1,2}(x_i, y_j) = p_{X,Y}(x_i, y_j).$$

Such a structure, (Ω, p, X, Y), is called a *statistical model* for the data. This is an example of *Kolmogorov's construction*. There might in fact be many such models; so the sample space constructed by Kolmogorov might not be the same as the original Ω one had in mind: it is a coarse graining of it, if X and Y do not separate the points of Ω. Indeed, we can construct Kolmogorov's model explicitly in the finite case, as follows.

Given $f_{X,Y}$ obeying (3.21), we define the Kolmogorov sample space to be set of distinct pairs of values (x_i, y_j) for which $f_{X,Y}$ is positive; we define the probability of the outcome (x_i, y_j) to be $f_{X,Y}(x_i, y_j)$, and the probability of any event by Eq. (3.1). Then one defines random variables X and Y in the obvious way, for example $X(x_i, y_j) := x_i$. Then it is an easy exercise to show that we have constructed a statistical model. It is the minimal model, in that it contains exactly the information that can be extracted by measuring just the variates X and Y. The sample points of Kolmogorov's model are exactly the joint level-sets of the two random variables of any other model of the same data; the variates X and Y separate the points of the sample space if and only if the model is Kolmogorov's model.

Naturally, one can generalise the construction and the conclusion to any other situation involving finitely many variates each taking finitely many values. The basic construction is that of the sample space, a point of which is an ordered list of values of the named variates. The construction includes any stochastic process with a finite number of time steps and the same finite set of values at each time, such as the Bernouilli process given above. I should say that this simple theorem is just a trivial case of Kolmogorov's work, which was created in order to treat cases infinitely many random variables labelled by continuous time; he showed that under suitable conditions, the consistency conditions allow one to construct countably additive measures on nice spaces, and measurable functions on this space whose distributions are the given ones.

Care is needed in human statistics, such as opinion polls, for which there might be no obvious underlying sample space. A related question has been raised by Accardi, and called the *chamelion effect*. Usually in analysing opinion polls, we assume that each individual has a well defined view on each question asked (agree, disagree, don't know, recorded as the values $1, -1, 0$). In this model, each question is a random variable taking three possible values. However, it is found that sometimes an answer on a questionnaire with more than one question can be influenced by what other questions are asked, and in what order. Then the sample space cannot be taken as the set of individuals; the answers to a given question are not well defined, so the questions are not random variables. Some examples have been worked out by Christensen [43]. He shows that in a model with four questions where the answer is contingent on who is asking the question, the correlations between the answers violate the Bell inequalities, calculated on the assumption that Ω is the set of people being questioned. These inequalities will be derived in Chap. (8). One should not, however, jump to the conclusion that such a system, violating Bell's theorem, needs the quantum treatment. For example, in [44] the authors ask a person from a random group to answer two psychological questions, A and B, in the order A, then B. Another random person is asked the questions B, then A. Another is asked just the question A, and another the question B. They analyse the results on the assumption that possible results of the questions, yes or no, make up the sample space. Thus they assume that $\Omega = \{(0,0), (0,1), (1,0), (1,1)\}$. They find that they get a contradiction with any probability on this space. This is interpreted as a quantum effect in the brain, instead of the need to consider a larger sample space. Indeed, the asking of question A often caused a different answer to question B to be given than if B is asked before A.

3.6 The Algebraic Theory of Gelfand

In the finite case, Kolmogorov's construction of a statistical model can be carried out starting from an information algebra $\mathcal{A}(X)$, rather than from Ω itself, if we are also provided with the expectation value of each random variable. When X separates points, we recover the original set-up (Ω, p). So what have we gained? In Gelfand's more abstract construction, we construct a statistical model without assuming that the algebra is derived from a probability model; indeed, this is one of the results proved. The version of the construction will be useful in our treatment of quantum theory, where there is no overall sample space.

First, some notation. We can distinguish the non-negative elements of $\mathcal{A}$ as those random variables X that are squares of others: $X = Y^2$. The expectation value $X \mapsto \mathbf{E}_p[X]$ defines a linear functional on this algebra; we shorten this functional to $p \cdot X$. Such a functional is *positive*: $p \cdot X \geq 0$ for all non-negative X. A positive linear functional is called a *weight*; it is called a *state* if $p \cdot 1 = 1$, as here.

We now give a construction of a commutative algebra, and a state on it, from a statistical model, and then give the converse, the construction of a statistical model from such an algebra and state. Thus, let (Ω, p) be a finite probability space, and let $\mathcal{A}$ be the vector space of random variables on (Ω, p). This is finite dimensional, spanned by the indicator functions of the outcomes. $\mathcal{A}$ has a natural product, the usual point-wise product of functions, thus: $XY(\omega) := X(\omega)Y(\omega)$. The product is obviously commutative, and has a unit (the sure function 1). Thus, $\mathcal{A}$ is a commutative real unital algebra of finite dimension, furnished with the state $X \mapsto p \cdot X$. We can extend it to include also every complex-valued random variable; this allows us to define the $^*-$ operation,

$$X^* := \overline{X},$$

meaning that this holds at every ω. This conjugation obviously obeys the relation

$$\|X^*X\| = \|X\|^2, \tag{3.22}$$

where $\|X\|$ is the norm

$$\|X\| := \sup_{\omega}\{|X(\omega)|\}.$$

Gelfand saw that Eq. (3.22) is essential in the analysis of $^*-$algebras, in that it enables a lot of calculations to be done. It is now called Gelfand's relation. Although far from obvious, Gelfand put his finger on the right axioms for the concept of an *abstract C^*-algebra*: it is a normed $^*-$algebra, complete in the topology defined by the norm, obeying Eq. (3.22).

Gelfand noticed that each point $\omega \in \Omega$ defines a *character* ρ_ω on the algebra; a character is a state, obeying the multiplication rule

$$\rho(XY) = \rho(X)\rho(Y). \tag{3.23}$$

One defines ρ_ω by evaluation of X:

$$X \mapsto \rho_\omega(X) := X(\omega).$$

We see that $\rho_\omega(X)$ is non-negative if X is non-negative, and is linear

$$\rho_\omega(\lambda X + Y) = (\lambda X + Y)(\omega) = \lambda X(\omega) + Y(\omega) = \lambda \rho_\omega(X) + \rho_\omega(Y);$$

Here, λ is any real number. We also see that $\rho_\omega(1) = 1(\omega) = 1$, so ρ_ω is a state. It is multiplicative, since

$$\rho_\omega(XY) = (XY)(\omega) = X(\omega)Y(\omega) = \rho_\omega(X)\rho_\omega(Y).$$

Thus each sample point gives us a character of the algebra. This is the key to proving the converse.

The converse theorem says (in finite dimensions) that any real unital commutative C^*-algebra $\mathcal{A}$ can be written as the set of random variables on a suitable minimal space, denoted by $\Omega(\mathcal{A})$. Moreover, any state ρ on the algebra defines a probability p on $\Omega(\mathcal{A})$, such that $\rho(X)$ of any element in the algebra is the expectation of the corresponding random variable in the measure p.

To outline the proof, we first give the sample space: $\Omega(\mathcal{A})$ consists of all characters ω of $\mathcal{A}$. Each element X of the algebra can be mapped to the random variable on $\Omega(\mathcal{A})$ given by $\omega \mapsto \omega(X)$. This mapping is an isomorphism of algebras; the proof of this is harder; how do we know that there are any characters? The proof starts by showing that the subsets of $\Omega(\mathcal{A})$, the events of the probability model, correspond to projections, and that the points of $\Omega(\mathcal{A})$, the characters, correspond to the minimal projections (those that are not sums of other projections). To show that we have an isomorphism then amounts to showing that the algebra is generated by its minimal projections. One way to prove this is to show that every X has a spectral representation, $X = \sum_i x_i P_i$, which is certainly true of matrix algebras. Gelfand showed that it also holds for any commutative C^*-algebra.

One then can check that a state ρ on the algebra provides us with a probability p on Ω giving the same means as ρ. In this way, each abelian algebra and state on it provides us with a statistical model.

We shall use this construction to formulate the modern version of the Copenhagen interpretation of quantum mechanics, which incorporates superselection rules and several observers.

Quantum Probability

"The Copenhagen interpretation *is* quantum mechanics."

R. Peierls.

We present the theory of quantum probability in the finite-dimensional case. We explain the effect of a measurement, as a completely positive stochastic map, and also discuss the case of two or more observers, which is called the quantum game. So this chapter, while entirely elementary, may still contain something new to the reader.

4.1 Algebras

A quantum theory must involve the dual concepts of algebra and state. The algebra will tell us the objects that we *measure*, and so will contain the observables of the theory. The states on the other hand tell us the probabilities that we get this or that value for the observable being measured. We start with algebra, of which the elements are a quite natural generalisation of random variables.

Heisenberg had the idea that we should represent an observable by a square matrix of complex numbers; this is instead of by a real function on phase space, as had been the case up till then. The set of all complex matrices has a natural conjugation, $*$: the (ij) entry of A^* is defined to be the complex conjugate of the (ji) entry of A. The map $A \mapsto A^*$, for all $n \times n$ matrices is called hermitian conjugation. It is anti-linear and its square is the identity map; this means that $(A^*)^* = A$ for any matrix A. A is said to be *self-adjoint*, or hermitian, if $A = A^*$. Matrices can be added and scaled (multiplied by complex numbers); the sum of two self-adjoint matrices is self-adjoint, and the real multiple of a self-adjoint matrix is also self-adjoint. Thus the self-adjoint matrices form a real vector space. Matrices in general do not commute, but can be multiplied to give other matrices. The product of non-commuting self-adjoint matrices is not self-adjoint, but inevitably enters the analysis. We must therefore include non-self-adjoint matrices in the theory. A vector space furnished with a multiplication is called an *algebra*, and if a conjugation is also specified, we have a $*$-algebra, provided that $(XY)^* = Y^*X^*$, which holds for matrices. Thus, Heisenberg was advocating the use of a $*$-algebra to formulate the new quantum theory.

The founders of quantum theory soon settled on the interpretation of this new representation of an observable. The possible realised values of an observable X are postulated to be the *spectrum* of the corresponding matrix, that is, the set of numbers x such that $X - xI$ is not invertible. The spectrum includes the eigenvalues, which are real if the matrix X is self-adjoint. It is natural, then, to consider a *complex matrix algebra* $\mathcal{M}$ as the arena of study, and to take the observables to be the self-adjoint elements of $\mathcal{M}$. An important case is the algebra $\mathcal{M}^n$, consisting of all complex $n \times n$ matrices. In this case, we may use the matrices to act on the complex vector space $\mathbf{C}^n$, consisting of column vectors, by left multiplication. A typical vector in $\mathbf{C}^n$ will be denoted by ψ, since it is the finite-dimensional analogue of Schrödinger's wave function. The spectrum of an operator then coincides with its eigenvalues; $X - xI$ is not invertible if and only if there is a non-zero vector ψ such that $X\psi - x\psi = 0$. $\mathbf{C}^n$ becomes a Hilbert space when we furnish it with the usual complex scalar product

$$\langle \psi, \varphi \rangle := \sum_{i=1}^{n} \overline{\psi}_i \varphi_i. \tag{4.1}$$

Recall that we use $\overline{\alpha}$ to denote the complex conjugate of the complex number α. A scalar product is positive-definite:

$$\langle \psi, \psi \rangle \geq 0 \tag{4.2}$$

for all $\psi \in \mathbf{C}^n$, with equality to 0 only when $\psi = 0$. Thus, the Hilbert space becomes a normed space with norm $\|\psi\| = \langle \psi, \psi \rangle^{1/2}$.

A (finite-dimensional) matrix U is called *unitary* if it leaves the scalar product invariant for all vectors φ, ψ, thus

$$\langle U\varphi, U\psi \rangle = \langle \phi, \psi \rangle.$$

Hermitian conjugation is related to complex conjugation in this scalar product:

$$\langle \psi, A\varphi \rangle = \langle A^*\psi, \varphi \rangle = \overline{\langle \varphi, A^*\psi \rangle}$$

for every matrix A. It is then a simple matter to show that a matrix is unitary if and only if $U^* = U^{-1}$.

The algebra $\mathcal{M}^n$ contains of course the hermitian [=self-adjoint] matrices, and among these are the *positive semidefinite* matrices, those self-adjoint matrices whose eigenvalues are ≥ 0. These are the 'positive' observables, and they define a convex positive cone inside $\mathcal{M}^n$; a convex positive cone (let us call it $\mathcal{C}$) is a subset such that

1. if $A \in \mathcal{C}$ and $B \in \mathcal{C}$, then $A + B \in \mathcal{C}$;
2. if $A \in \mathcal{C}$, then $\lambda A \in \mathcal{C}$ for all $\lambda \geq 0$;
3. if $A \in \mathcal{C}$ then $-A \notin \mathcal{C}$, unless $A = 0$.

It can be shown that the set of positive observables obeys these axioms. We shall now use $\mathcal{C}$ for this set. One can show that any matrix C in $\mathcal{C}$ can be written as $C = A^* A$ for some $A \in \mathcal{M}^n$.

Another important special case of a matrix algebra is the set $\mathcal{D}$ of all real diagonal $n \times n$ matrices. The eigenvalues of a diagonal matrix are the diagonal entries, and the eigenvectors are the elements of the natural basis in $\mathbf{C}^n$. That is, the elements

$$(1,0,0,\ldots,0), \quad (0,1,0,\ldots,0), \ldots, (0,0,\ldots,0,1).$$

$\mathcal{D}$ is a commutative [= abelian] algebra of dimension n over the real field. It can be regarded as the set of random variables on the sample space $\Omega_n = \{1,2,\ldots,n\}$. To see this, let $D \in \mathcal{D}$ be such a diagonal matrix, and define the function X_D on Ω_n by

$$X_D(\omega) := D_{\omega,\omega}, \qquad \text{for } \omega = 1,2,\ldots,n. \qquad (4.3)$$

Thus, X_D is a random variable on the sample-space Ω_n, whose possible values are the eigenvalues of D. Moreover, the algebraic structure of $\mathcal{D}$ is isomorphic to that of all random variables on Ω_n. This means that the map $D \mapsto X_D$ is bijective, and that if D and K are in $\mathcal{D}$, then

$$X_{\alpha D + \beta K} = \alpha X_D + \beta X_K \qquad \text{for all real } \alpha, \beta; \qquad (4.4)$$
$$X_{DK} = X_D X_K \qquad (4.5)$$

as functions on Ω; this follows from the definitions of product of diagonal matrices, and product of functions. Since $\mathcal{D}$ is a subalgebra of $\mathcal{M}^n$, it is natural to interpret quantum mechanics as a generalization of classical probability, rather than as a new mechanics. The analogue of a sure random variable is a multiple of the identity; so to provide a generalisation of probability theory, the chosen algebra $\mathcal{M}$ should be *unital*, that is, it should contain the identity, denoted I.

If $X \in \mathcal{M}^n$, and U is unitary, we say that $Y = U^{-1} X U$ is the unitary transform of X by U. We also say that Y is the conjugation of X by U. If so, then X and Y have the same eigenvalues, and the corresponding eigenvectors are related: multiply the eigenvector of X by U to get the eigenvector of Y, corresponding to the same eigenvalue. When we transform all the matrices of $\mathcal{M}^n$ by the same unitary matrix U, we get a map $\mathcal{M}^n \to \mathcal{M}^n$ in which all sums, scalar multiples, and products are preserved; thus the map is an algebraic isomorphism. It also preserves conjugation, which means that $U^{-1} X^* U = (U^{-1} X U)^*$ for all matrices X. Thus, a unitary transform is a *-isomorphism of the *-algebra of complex matrices. The unitary transform of a self-adjoint matrix is always self-adjoint.

The Heisenberg dynamics, which replaces Newton's laws of motion, is determined by a continuous one-parameter unitary group of operators, $\{U(t) : -\infty < t < \infty\}$; that is, for each t, $U(t)$ is unitary and is a continuous operator-function of t; it obeys

1. $U(0) = I$
2. $U(s)U(t) = U(s+t)$ for all $s, t \in \mathbf{R}$

These act on the observables by conjugation, so the observables become time-dependent: $X(t) := U(t)^{-1}X(0)U(t)$. If H is any self-adjoint matrix, then

$$U(t) := \exp(-iHt) := 1 - iHt + i^2 H^2 t^2/2! + \ldots$$

defines a convergent series of matrices; the sum is continuous in t and obeys (1) and (2). In fact, $U(t)$ is differentiable, and

$$\frac{dU(t)}{dt} = -iHU(t)$$

holds. In finite dimensions, any continuous family obeying (1) and (2) has this form for some self-adjoint matrix H, equal to $iU(t)^{-1}\frac{dU(t)}{dt}$ for all t. With care in the concept of self-adjointness, the same result also holds in infinite dimensions. When $U(t)$ is the time-translation operator of the theory, the generator of the group, the operator H, is called the *Hamiltonian* of the system.

Here and also in the rest of the book, we have chosen units so that $\hbar$, Planck's constant$/(2\pi)$, is 1.

Using these ideas, but using infinite dimensional matrices, Pauli was able to derive the energy-levels of the Bohr atom, just a few months before Schrödinger did the same using his wave mechanics.

4.2 The Schrödinger Picture

The dynamics described above, in which the observables evolve under transformation by a one-parameter unitary group, is called the *Heisenberg picture* of the dynamics. Just as Heisenberg emphasised the observables of a system, so Schrödinger emphasised the states, which he took to be wave-functions. In finite dimensions, these are the complex column vectors $\psi \in \mathbf{C}^n$. These evolve in time according to the Schrödinger equation, while the observables do not change with time. This is called the Schrödinger picture, in which the time-evolution of Heisenberg is replaced by the action on wave-functions, $\psi \mapsto U(t)\psi(0) := \psi(t)$. The dynamics is determined by H, in that the wave-function at time t is determined by the unique solution of Schrödinger's equation

$$i\frac{d\psi}{dt} = HU(t)\psi(0) = H\psi(t). \tag{4.6}$$

These two pictures are physically equivalent; this means the following. Whereas it is understood that any single measurement of the observable represented in quantum theory by the self-adjoint operator X will be one of the eigenvalues of X, we do not always get the same eigenvalue in a run of measurements on a beam of identically prepared samples. Instead, we get a statistical scatter of eigenvalues, unless the state is an eigenstate. We interpret $\langle \psi, X\psi \rangle$ as the expectation value of the observable X when the wave-function is ψ. To fix the probabilistic meaning, it is enough to give the expectation of every observable. To be consistent, the mean of the unit operator must be 1, which leads to the requirement that the wave-functions are normalised: $\langle \psi, \psi \rangle = 1$. Such vectors are also called unit vectors.

Experiments with the absorption and transmission of polarised light led to the interpretation of the wave function in terms of probability: if the (normalised) wave-function ψ falls on a Nicol prism that only transmits the unit vector φ, then the probability of transmission is $|\langle \varphi, \psi \rangle|^2$. The square in this formula is related to the fact, known from classical Maxwell theory, that the energy of an electromagnetic wave is proportional to the square of the fields. The time-evolution of the expectation values is the same, whether we use the Heisenberg or the Schrödinger picture:

$$\langle \psi, X(t)\psi \rangle = \langle \psi, U(t)^{-1}XU(t)\psi \rangle$$
$$= \langle U(t)\psi, U(t)U(t)^{-1}XU(t)\psi \rangle = \langle \psi(t), X\psi(t) \rangle.$$

This relation shows the equivalence of both pictures.

It follows from the unitarity of U that if $\psi(0)$ is normalised, so is $\psi(t)$ for all t. The expectation of any matrix is unchanged if we change ψ by multiplying it by a complex number of unit modulus. Since any observable property of a state is determined by the expectations of all observables, the vectors ψ and $e^{i\theta}\psi$ are indistinguishable whatever the (real) value of θ. Thus, the wave-function is ambiguous up to a phase; instead of defining the state as the wave function, the essence of the *state* is the collection of expectation values. Modern quantum theory requires us to consider a generalisation of Schrödinger's concept of state, to mixed states. We get a mixture of two states from a source that selects one vector ψ with probability λ and a different vector φ with probability $1 - \lambda$. This can also be regarded as a state, giving us a well-defined expectation for each observable, X, namely, it gives us the weighted average of the means in the two states ψ and φ:

$$X \mapsto \lambda \langle \psi, X\psi \rangle + (1 - \lambda)\langle \varphi, X\varphi \rangle. \tag{4.7}$$

This is called the incoherent mixture of ψ and φ. It should be contrasted with the *coherent* mixture, given by the vector $\xi = c(\lambda\psi + (1 - \lambda)\varphi)$, where c is chosen so that the sum is a unit vector. It is clear that the expectation $X \mapsto \langle \xi, X\xi \rangle$ differs from (4.7) for some $X \in \mathcal{M}^n$. This difference is caused by the presence in the coherent mixture of the cross-terms

$$|c|^2\lambda(1 - \lambda)[\langle \psi, X\varphi \rangle + \langle \varphi, X\psi \rangle].$$

This is zero (if ψ and φ are orthogonal) only if X has no non-zero off-diagonal matrix elements in the basis formed by these vectors. Thus, the coherent and incoherent mixtures are physically distinguishable, if we allow all hermitian operators to be observables. We shall offer a small generalisation of this idea when we discuss *superselection rules.*

4.3 State Space

The set of all states is best studied as the positive part of the (linear) dual of the algebra of observables.

The linear dual of a vector space is the set of linear functionals on the space. For a complex vector space such as $\mathcal{M}^n$, the dual is the set of complex-linear functionals. [Functional is an old-fashioned word for function taking real or complex values]. The set of states of a C^*-algebra $\mathcal{M}$ consists of linear functionals ρ such that

1. ρ is *positive*: that is, $\rho(A^*A) \geq 0$ for all matrices A.
2. ρ is normalised: $\rho(1) = 1$.

If the second condition is dropped, the functional is called a *weight*. If the dimension is infinite, and one wants to drop (2), then ρ is called a *finite weight*. It is easy to prove that if $X = X^*$, then for any state ρ, $\rho(X)$ is real. We interpret $\rho(X)$ as the expectation value of the observable X when the system is in state ρ. Any normalised column vector ψ defines a state, by the functional $X \mapsto \langle \psi, X\psi \rangle$. Each random mixture of two vectors, as in Eq. (4.7), defines a state. In each case, the interpretation of the state $X \mapsto \rho(X)$, as given by Born, is as the expectation value of the observable X. This is thus a reasonable interpretation of the phrase, the state of the system is ρ, also for any more general ρ.

Further examples of states on a matrix algebra are given by a density matrix Δ; that is, Δ is positive definite and its trace is 1. Then the corresponding state is ρ_Δ, which is the functional

$$A \mapsto \rho_\Delta(A) := \mathrm{Tr}\,(\Delta A).$$

Recall that the trace of a matrix X, written $\mathrm{Tr}\,X$, is the sum of its diagonal values. The trace of a product XY is the same as the trace of YX. From this, we can conclude that the trace has the cyclic property: $\mathrm{Tr}\,(XY \ldots Z) = \mathrm{Tr}\,(Y \ldots ZX)$. It follows that the trace is invariant under similarity transformations: $\mathrm{Tr}\,(S^{-1}XS) = \mathrm{Tr}\,X$ for any invertible matrix S. In particular, the trace is unitary invariant.

A special class of states is given by a diagonal matrix D_1 with 1 in one place, say $D_{11} = 1$, the rest being zeros. This operator is the projection onto the vector ψ_1 say, in $\mathbf{C}^n$. A projection is any operator P such that $P^2 = P$. It follows that the eigenvalues of P are zero and 1. For, the eigenvalues of a matrix obey the same polynomial equations as the matrix itself (Caley's theorem). If P is also self-adjoint, it is called an orthogonal projection. We shall call it a projector, for short. A projector defines a subspace of the Hilbert space $\mathbf{C}^n$, namely, the *range* of the operator, which is the set of vectors of the form $P\psi$ as ψ roams over $\mathbf{C}^n$. If the range is one-dimensional, it is spanned by any non-zero vector in it, say ψ_1. In the basis in which ψ_1 is the first vector, the operator P is diagonal, with 1 in the position $(1,1)$ and zeros elsewhere. Then P is a density operator, defining a state. Such states are called *vector states*. The reason for the name is that it defines the same expectation value functional as that given, in Schrödinger's theory, by any normalised wavefunction, say ψ_1, spanning its range:

$$\mathrm{Tr}\,[D_1 A] = \langle \psi_1, A\psi_1 \rangle. \qquad (4.8)$$

It is not difficult to show that mixtures of vector states give us the most general density matrix, and that any state ρ on a matrix C^*-algebra is given by a density matrix, which we denote by Δ_ρ.

4.4 The Spectral Theorem

Any self-adjoint operator A can be written in terms of its eigenvalues $\{\lambda_i\}$ and the corresponding projectors:

$$A = \sum_i \lambda_i P_i \qquad (= \text{the spectral resolution of } A).$$

Let ρ be a state on $\mathcal{M}^n$. The functional $A \mapsto \rho(A^j) = m_j(A)$ for any power of A, has the interpretation as the moments of a random variable, since the real numbers m_j, obey all the properties of a moment sequence (3.14) in classical probability. Namely, for any n, and any complex $\alpha_1, \ldots, \alpha_n$, we have

$$\rho\left(\left(\sum_{i=1}^n \alpha_i A^i\right)^* \left(\sum_{j=1}^n \alpha_j A^j\right)\right) = \sum_{i,j} \overline{\alpha}_i \alpha_j \rho(A^i A^j) = \sum_{i,j} \overline{\alpha}_i \alpha_j m_{i+j}(A) \geq 0,$$

$$(4.9)$$

giving Eq. (3.16). Thus, the sequence $\{m_i\}$ is the moment sequence of a random variable, by the sufficiency of (3.16). The interpretation of $\rho(A^j)$ as the mean of A^j leads to the construction of a probability distribution for the values of A. This turns out to imply that only the eigenvalues $\{\lambda_i\}_{i=1,\ldots n}$ are possible values, which was originally given as a separate postulate. For the probability of such a value, it gives the following:

$$\mathrm{Prob}.\{A = \lambda_i\} = \mathrm{Tr}\,[P_i \Delta_\rho P_i] \qquad (4.10)$$

where $A = \sum_i \lambda_i P_i$ is the spectral resolution of A. Suppose that all eigenvalues are different; then P_i is a density operator and the probability in the state P_i that A has the value λ_i is $\mathrm{Tr}\,(P_i P_i P_i)$, which is 1 as it should be.

The states of $\mathcal{M}^n$ form a convex set, when we use the linearity of the dual space to define the affine structure [which is how we define the mixture of two states]. A state ρ is said to be an extreme point if it cannot be written as a mixture, $\rho = \lambda \rho_1 + (1 - \lambda)\rho_2$, with $\rho_1 \neq \rho_2$ and $0 < \lambda < 1$. The extreme states on $\mathcal{M}^n$ are called pure, and are in fact, the vector states. Every density operator can be written as a convex mixture of pure states, so we might interpret the state as arising by the mixing of pure states with various weights. I say, might interpret, because in general, the decomposition of a state into pure states is not unique. The same ρ can usually be written as the statistical mixture of pure states, or other mixed states, in many ways. So, the interpretation is ambiguous. It is a fact, however, that we can produce a mixed state by physically mixing beams of particles in pure states; and, moreover, two such mixtures composed of different sets of pure states are physically indistinguishable if they lead to the same density matrix: the details of how the state was made are lost in the mixture. The density matrix is all there is. To see this, just note that two states are taken to be equal if they define the same linear functional on the algebra; they thus give the same results for any probability distribution of an observable. A physicist might arrive at a mixed state of a system S without mixing systems; he can start with a pure state, then allow S to interact with another system, and then limit the state (which is to be regarded as a function on $\mathcal{M}$) to the observables of S. The resulting mixed state cannot be said to have been made by physically mixing the pure states which make it up.

This ambiguity in decomposing a mixture into pure states does not occur in classical probability; a probability p on Ω can be written as the sum of the indicator function at each point ω_i, weighted with $p(\omega_i)$:

$$p(\omega) = \sum_i p(\omega_i)\delta(\omega, \omega_i).$$

This decomposition is obviously unique. Because of this, we say that classical state space is a *simplex*. The quantum state space, then, is not a simplex.

As an example, consider the matrix algebra $\mathcal{M}^2$, and its dual; by what we said above, this is isomorphic to the space spanned by two-by-two density matrices, which is the same space, $\mathcal{M}^2$. As a vector space, $\mathcal{M}^2$ has dimension 4, and a basis is provided by the unit matrix and the three Pauli spin matrices:

$$\sigma_x := \begin{pmatrix} 0 & 1 \\ 1 & 0 \end{pmatrix}; \qquad \sigma_y := \begin{pmatrix} 0 & -i \\ i & 0 \end{pmatrix}; \qquad \sigma_z := \begin{pmatrix} 1 & 0 \\ 0 & -1 \end{pmatrix}. \quad (4.11)$$

It happens that these operators are twice times the generators of rotations, which represent infinitesimal rotations of the components of spin $1/2$ around the three axes, respectively. They do not commute, but satisfy the relations

$$\sigma_j \sigma_k + \sigma_k \sigma_j = 0, \qquad\qquad \sigma_j \sigma_k = i\sigma_\ell, \qquad\qquad \sigma_m^2 = I,$$

j, k, ℓ being cyclic permutations of x, y, z and m being any component. The matrix σ_z is diagonal, and its eigenvalues can be read out as ± 1. The eigenvectors are $(1,0)$ and $(0,1)$. Let $P_z(\pm)$ be the projections onto these two vectors. They are complementary projections, in that $P_z(+)+P_z(-) = I$. The spectral theorem for σ_z then reads

$$\sigma_z = 1.P_z(+) + (-1).P_z(-) = P_z(+) - P_z(-).$$

It is clear that σ_z commutes with $P_z(+)$ and with $P_z(-)$, as it should: any self-adjoint operator commutes with its spectral projectors. Likewise, σ_x and σ_y have eigenvalues ± 1, and spectral projectors $P_x(+), P_x(-)$ and $P_y(+), P_y(-)$ respectively. They are also complementary projectors, so we have the three relations

$$I = P_x(+) + P_x(-) = P_y(+) + P_y(-) = P_z(+) + P_z(-).$$

The matrix $P_z(+)$ is a density matrix, representing a spin aligned in the $+z$ direction; the state is pure, and is said to be completely polarised; Similarly $P_z(-)$ is the density matrix of a spin completely polarised in the $-z$ direction. Similarly we interpret $P_x(\pm)$ and $P_y(\pm)$. The density matrix $\frac{1}{2}I$ is thus the equal mixture of $P_x(+)$ and $P_x(-)$, and represents a state with completely unpolarised spin. It can also be written as the equal mixture of $P_y(\pm)$, or of $P_z(\pm)$, showing that these mixtures give the same density matrix and so there is no physical difference between them. This is possible here, since the algebra generated by all the spins, $\sigma_x, \sigma_y, \sigma_z$, is M^2, a *non-abelian* algebra.

4.5 Tensor Products

We can combine two logically independent systems into one combined system by using the tensor product. This can be done in the Schrödinger picture using states, or in the Heisenberg picture using algebras. We must distinguish three different senses in which the word 'independent' is used in physics. These are logical independence, dynamical independence and statistical independence.

In logic, we say that two propositions P and Q are logically independent if any of the four propositions 'P is true and Q is true', 'P is true and Q is false', 'P is false and Q is true', and 'P is false and Q is false, could be true. Giving truth values $1, 0$ to P and Q, the four propositions are elements of the product set $\{0,1\} \times \{0,1\}$. More generally, we can consider a degree of freedom in classical mechanics as being described by a phase space consisting of points $(q_1, p_1) \in \Gamma_1$. We would then say that another degree of freedom $(q_2, p_2) \in \Gamma_2$ is an independent degree of freedom if the combined system can be described by the space $\Gamma_1 \times \Gamma_2$; the values allowed for each variable do not depend on the values taken by the other. This is often loosely expressed by saying that there is no functional relation between the coordinates. The quantum analogue of this concept of independence is that of tensor product. We know that a particle in one dimension is described by wave-functions depending on one variable, such as $\psi(q_1)$, which form the space $L^2(\mathbf{R})$. In two dimensions, we use wave-functions $\psi(q_1, q_2)$ of two independent variables. The space of wave functions is then $L^2(\mathbf{R}^2)$, which is isomorphic to the 'tensor product' of two copies of $L^2(\mathbf{R})$. We now describe the tensor product of vector spaces, at least in finite dimensions.

Suppose that $\mathcal{H}$ and $\mathcal{K}$ be complex vector spaces of dimension m, n respectively; let $\{\psi_1, \ldots, \psi_m\}$ be a basis of $\mathcal{H}$ and $\{\varphi_1, \ldots, \varphi_n\}$ a basis in $\mathcal{K}$. Then there are $m \times n$ symbols $\{\psi_i \otimes \varphi_j\}$, which are taken to be the basis of a vector space, the tensor product $\mathcal{H} \otimes \mathcal{K}$; its dimension is of course mn. Elements of the tensor product are thus symbols of the form $\sum_{ij} c_{ij}(\psi_i \otimes \varphi_j)$. The following combinations define equivalence among tensors:

1. $\lambda(\psi \otimes \varphi) \equiv (\lambda\psi) \otimes \varphi \equiv \psi \otimes (\lambda\varphi)$ for all $\lambda \in \mathbf{C}$
2. $(\psi + \psi') \otimes \varphi \equiv \psi \otimes \varphi + \psi' \otimes \varphi$
3. The same as (2) but with the sum in the second factor.

It can be shown that the tensor product is essentially independent of the choice of bases used in its construction.

If both $\mathcal{H}$ and $\mathcal{K}$ are Hilbert spaces of finite dimension, they are still vector spaces, and we define their tensor product as a vector space as above. However, we have a scalar product defined in each space, and this allows us to provide the tensor product with a scalar product, so that it becomes a Hilbert space. To do this, we choose an orthogonal basis in each space, and take the product basis in $\mathcal{H} \otimes \mathcal{K}$ to be an orthogonal basis. This, with the linear rules of tensor product, can be shown to determine a unique scalar product in the tensor product, independent of which orthogonal bases are used in its definition. In infinite dimensions, the construction works for any finite-dimensional subspaces of $\mathcal{H}$ and $\mathcal{K}$; and so to define the tensor product of two Hilbert spaces, $\mathcal{H}$ and $\mathcal{K}$ we can complete the union of all these constructions in the norm topology, to get a unique Hilbert space, the tensor product $\mathcal{H} \otimes \mathcal{K}$.

We can also form the tensor product of algebras. Let $\mathcal{A}$ and $\mathcal{B}$ be *-algebras of finite dimension. In particular, they are vector spaces. We can thus form the tensor product of these vector spaces, to get another vector space, in which products $A \otimes B$ give, by definition, a total set (they span the space). We can then define the product of two such: $A \otimes B \times C \otimes D = (AC) \otimes (BD)$, and this can be extended to all elements of the span by linearity. We can take the tensor product of the algebras of observables of two systems to define the observable algebra of the situation in which both systems are present. The sub-algebra $\mathcal{A} \otimes I$ can be identified with the algebra $\mathcal{A}$, and the sub-algebra $I \otimes \mathcal{B}$ can be identified with the algebra $\mathcal{B}$. This is the meaning of logical independence of two systems.

Now we turn to dynamical independence. The construction, the tensor product, does not mention time-evolution. The two systems put together in this way, might or might not be dynamically independent. It is better to use the expression, non-interacting, as the word independent is becoming over-used. Let $\mathcal{H}$ and $\mathcal{K}$ be the Hilbert spaces of two logically independent systems, with time evolution given in the Schrödinger picture by $U(t)$ and $V(t)$

respectively. Suppose that the combined system evolves with the unitary group $U(t) \otimes V(t)$; then we say that they are non-interacting, or dynamically independent. This is the same as saying that the Hamiltonian of the combined system is the sum of the Hamiltonians of each: there is no potential energy of interaction. Suppose we write $U(t) = e^{-iH_1 t}$ and $V(t) = e^{-iH_2 t}$; that is, suppose that H_1 and H_2 are the generators of $U(t)$ and $V(t)$. We see that the generator of the time-evolution $U(t) \otimes V(t)$ is then $H = H_1 \otimes I + I \otimes H_2$. We might instead of this take an interaction, $H + V$, where V is any operator on $\mathcal{H} \otimes \mathcal{K}$. Thus, we see that the logically independent systems might, or might not, be interacting.

The third use of the term independent concerns statistical independence. This is a different notion from the other two; consider first the classical case, when the algebra of observables is abelian, and we have a state, ρ. Then we say that two observables X, Y are statistically independent if, in the associated statistical model, all the events defined by the level sets of X are independent of all the events defined by the level sets of Y. Similarly, we define the statistical independence of any number of commuting observables. As it happens, sometimes statistical independence arises in a tensor product of subsystems. For, let $\mathcal{A}$ and $\mathcal{B}$ be logically independent algebras of observables, not necessarily abelian. Let states ρ and σ respectively, with density operators $\Delta_\rho, \Delta_\sigma$, be given. Then the algebra $\mathcal{A} \otimes \mathcal{B}$ can be furnished with the state given by the density operator $\Delta_\rho \otimes \Delta_\sigma$. Then in this state, any observable of the form $A \otimes I$ is statistically independent of any observable of the form $I \otimes B$. However, there are many states on the product algebra that are not product states; then observables in $\mathcal{A}$ could well be correlated (in this state) with an observable in $\mathcal{B}$, even if, physically, the two algebras concern measurements in regions of space separated by a huge distance. This can happen in quantum theory, in which the algebras are not abelian, as well as when they are abelian. Given a tensor product $\mathcal{C} = \mathcal{A} \otimes \mathcal{B}$ of (abelian or non-abelian) algebras $\mathcal{A}$ and $\mathcal{B}$, we say that $\mathcal{A}$ is statistically independent of $\mathcal{B}$ if and only if the state is a product state, $\Delta_\rho \otimes \Delta_\sigma$. Clearly, a non-product state is possible even although $\mathcal{A}$ and $\mathcal{B}$ are logically independent, as here; it can also happen when $\mathcal{A}$ and $\mathcal{B}$ are dynamically independent. It is not a quantum phenomenon, as it can occur when both $\mathcal{A}$ and $\mathcal{B}$ are abelian.

Suppose that $\dim \mathcal{A}$ and $\dim \mathcal{B}$ are each ≥ 2. Then the span of the product states of the form $\rho \otimes I$ and $I \otimes \sigma$ do not span $\mathcal{C}$; for the vector $\psi_1 \otimes \psi_2 - \psi_2 \otimes \psi_1$ is not in it, when $\mathcal{A} = \mathcal{B}$ and $\psi_1 \neq \psi_2$. This fact will come into play in the Bell inequality.

4.6 Superselection Rules

Wigner noticed as early as 1931, in his book on atomic spectroscopy [172], that not every self-adjoint operator can correspond to an observable. In quantum theory the Hilbert subspace containing the wave-functions of particles of spin

0 (or any integer spin) are said to form a coherent subspace, as are the wave-functions of particles of spin $1/2$ (or any half-odd-integer). Recall that we take $\hbar = 1$. The modern word for coherent subspace is *sector*. Wigner's classic and convincing argument makes use of the Pauli spin matrices. These are hermitian, and the exponential

$$R_x(\theta) := \exp\{i\theta\sigma_x/2\}$$

is unitary; it is defined by a convergent infinite power series of matrices. Its action on a two-component wave function gives the state representing the system rotated by the angle θ about the x-axis. We say that $\sigma_x/2$ is the generator of the rotation group about the x-axis. Similarly, $\sigma_y/2$ and $\sigma_z/2$ generate rotations about the $y-$ and $z-$axes. We find, however, that the rotation matrices R do not quite represent the group of rotations. The product law holds only up to some phase (complex number of unit modulus); the ambiguity in the *phase* is then a number in $[0, 2\pi)$; for example

$$R_x(\theta_1)R_x(\theta_2) = \pm R_x\left((\theta_1 + \theta_2)\mathbf{mod}(2\pi)\right).$$

The minus sign arises if $\theta_1 + \theta_2 > 2\pi$. Recall that the effect of adding the symbol $\mathbf{mod}(2\pi)$ is to subtract 2π from the sum $\theta_1 + \theta_2$ if this should be greater than 2π, and not to change anything otherwise.

   More generally, we may represent a rotation by an angle $|\boldsymbol{\theta}|$ about an axis in the direction of $\boldsymbol{\theta}$ by the unitary matrix

$$R(\boldsymbol{\theta}) = e^{i\boldsymbol{\theta} \cdot \boldsymbol{\sigma}/2},$$

where $\boldsymbol{\theta} \cdot \boldsymbol{\sigma} = \sum_{j=1}^{3} \theta_j\sigma_j$. Since the phase of a wave-function makes no difference, we might try to change the Rs by adjusting the phase carefully for each θ. Such a change, using a complex function $b(\boldsymbol{\theta})$ of modulus 1,

$$R(\boldsymbol{\theta}) \rightarrow R'(\boldsymbol{\theta}) := b(\boldsymbol{\theta})R(\boldsymbol{\theta}),$$

is called an allowable phase change. Wigner showed that one could choose b so that the R's nearly obey the group law, which here is

$$\boldsymbol{\theta}_1 \circ \boldsymbol{\theta}_2 = \boldsymbol{\theta}_3, \text{ the rotation group.} \tag{4.12}$$

In fact, we can get for the case of spin $1/2, 3/2, \ldots$:

$$R(\boldsymbol{\theta}_1)R(\boldsymbol{\theta}_2) = \pm R(\boldsymbol{\theta}_3) \tag{4.13}$$

for any rotations; then a complete rotation by 2π changes the sign of the wave function. For particles of integer spin Wigner showed that we can remove the phase $\pm$, so that the rotation by the angle 2π becomes the identity operator.

   A vector ψ with non-zero components in both the sector of spin zero and the sector of spin $1/2$, say $\psi = \psi_0 + \psi_{1/2}$ is not physically realizable. For, under a rotation by 2π about any axis, such a vector changes to $\psi_0 - \psi_{1/2}$, which differs from ψ by more than an overall phase. Wigner then argued that for a system that was invariant under a rotation, all physically observable properties of a system must be unchanged by such a rotation, which is the identity of the group. In particular, the projection operator onto vectors such as ψ is not invariant, and so cannot be observable: ψ cannot be realised in Nature. We say that a superselection rule holds between subspaces if we are not allowed to mix them *coherently*. The modern version of superselection sectors rewrites the theory in terms of operators; we do not prohibit linear sum of such vectors (such a move would cause inconvenience) but achieve the same result: the linear sum behaves as an incoherent mixture. To do this, limit the observables to those operators that map each sector into itself: an observable cannot have non-zero matrix elements between the two sectors. The algebra of observables $\mathcal{M}$ then does not contain all operators on the Hilbert space. With this limitation, the forbidden wave-function $2^{-1/2}(\psi_0 + \psi_{1/2})$ (a perfectly good vector in the orthogonal direct sum Hilbert space of the two sectors) is allowed, but defines the *same state* on $\mathcal{M}$ as the incoherent mixture $\frac{1}{2}(\rho_0 + \rho_{1/2})$. This is the mixture of the two linear functionals by purely classical probability. Thus was born the *univalence* superselection rule. The name arises because the representations of the rotation group with spin $1/2, 3/2, 5/2 \ldots$ are not true representations, but are projective representations, like the case of spin $1/2$. That is, the operators implementing rotations on the Hilbert space obey the group law only up to a phase, called a cocycle of the group. Since two vectors that differ by an overall phase give rise to the same density operator, this is an allowed realisation of the symmetry group. Thus, Wigner showed that the cocycle can be adjusted by allowed phase changes, until it takes the values ± 1. Then these representations became known as *two-valued* representations, because the cocycle can take two values. Particles of spin $0, 1, 2, \ldots$ do not need a cocycle in their representations of $SO(3)$; they are true representation with trivial cocycle, equal to 1 after suitable allowed changes of phase. These became known as *one-valued* representations. To maintain invariance under $SO(3)$, we cannot coherently mix one-valued and two-valued representations. Without this rule, there would be no electrons, protons or neutrons in the same universe as photons. The univalence rule remained private to Wigner and his students for twenty years, until G. C. Wick heard about it from Wigner. Wick, Wightman and Wigner applied the idea to the measurablility of phases in time-reversal and parity operators, resulting in the paper [167]. From a modern viewpoint, this paper assumes too much, in that parity and time-reversal are exact symmetries; see Wightman [171].

In Wigner's theory, only the *overall* phase of the vector in unobservable; the relative phase between different vectors is observable, at least in principle. For example, suppose that the state is given by the vector $\psi = 2^{-1/2}[\psi_1 + \psi_2]$ and we compare this with the state given by the vector $\phi(\theta) = 2^{-1/2}[\psi_1 + e^{i\theta}\psi_2]$. These predict different expectation-values for all operators which link the two subspaces spanned by ψ_1 and ψ_2. This idea has been put to the test by Rauch *et al.* [140], who confirmed that a rotation of a spinor wave-function by 2π changes its sign. These authors started with a beam of unpolarised neutrons of a well-defined velocity. This beam is described by a linear sum of states of a single particle. They split the beam using Rauch's neutron beam-splitter, getting two spatially distinct parts of the beam. One part of this beam, ψ_0 is untouched, while the other part, ψ_H passes through a nearly constant magnetic field, **H**. The neutron in the magnetic field undergoes Larmor rotation through an angle α, which is calculated by the authors. According to theory, the part of ψ_H with spin along **H** changes by a phase $e^{i\alpha/2}$, and the part of ψ_H with spin along $-$**H** changes by a phase $e^{-i\alpha/2}$. The two parts of the beam, ψ_0 and the modified ψ_H, were brought together and the resulting interference was analysed. It was found that the observed pattern gave the same answer for the values α and $\alpha + 4\pi$, as predicted by the theory. One part of the system, described by the state ψ_H, is not isolated from the other part, described by the state ψ_0; the result does not violate the univalence superselection rule. To do this, one should rotate both parts, $\psi = \psi_0 + \psi_H$, by the angle 2π, not just one part, to get the vector $-(\psi_0 + \psi_H)$, and show a difference with the state given by ψ; of course, there would be no difference, and this was not the purpose of the experiment.

The analysis of representations of the Galilean group in quantum theory led to another superselection rule, this time involving the mass. In 1948, Bargmann [13] showed that the solutions to the Schrödinger equation transform under the Galilean group as a projective representation. There are some more (true) representations, but they do not occur in quantum theory. The cocycles found by Bargmann were phases containing a factor e^{imt}, where m is a number identified with the mass of the particle. For several particles, it is the total mass; and, systems with different total masses cannot be coherently superposed (without violating Galilean invariance). Not only does this explain non-relativistic mass by group theory, it gives us another superselection rule (Bargmann's).

When we have a superselection rule, the dynamics cannot mix states across the coherent subspaces. Thus, a fermion cannot turn into a boson (without violating rotation symmetry), and the total mass in any Schrödinger equation must be conserved (unless Galilean invariance is also violated). This enables one to say, on sight, that non-relativistic models of quantum field theory, that do not conserve the total mass, violate Galilean invariance. We may wonder why Nature does not use a *true* representation of the Galilean group, rather than the projective representations used in the Schrödinger equation; the true representations were all found by Mackey, in his famous work [113] on group representations. The answer seems to be that the Galilean group is an approximation to the Poincaré group, $\mathcal{P}$. We require a representation of the Galilean group that approximates the representations of $\mathcal{P}$. Now, in the Galilean group, the transformations of Galileo, the boosts, which change the Galilean frame of reference, commute with the space translations; in $\mathcal{P}$, the boosts do not commute with space translations. Thus, Nature solves the approximation problem by choosing only representations in which boosts do not commute with translations; these are the projective representations found by Bargmann. The generator of the boost turns out to be the position operator. Thus, Heisenberg's commutation relation has a relativistic origin. In classical physics, we also get a multiplier for the Galilean group, acting on the observables, for the same reason.

If the coherent subspaces of a theory with superselection rules are labelled by a number, then this number must be conserved in time whatever the dynamics. More is needed, as it requires that observables also commute with the superselection operators. Some conservation laws were called selection rules in the old quantum theory. Hence the term superselection rule was coined for the corresponding conservation law. The inability to measure the phase between the coherent subspaces is absolute; it is not something that appears because of the absence of classical measuring devices in stable coherent mixtures.

It is believed that there is a superselection rule between states of different charge; this follows from the gauge invariance of electrodynamics: if we add the gradient of any function Λ to the magnetic four-potential, A^μ, we do not change the electric or magnetic fields. This invariance is postulated to hold in the quantum version. This superselection rule leads to the absolute conservation of electric charge. Naturally, by 'charge' we must be talking about the total charge; if we split the charge into an integral over a charge density, then we can measure the relative phase of the wave-function localised in two different regions. A proof of the superselection rules for states of different charge was started by Haag and Swieca [80]. It is based on Gauss's law for the electric field due to a charge. These authors assumed smoothness properties that do not hold in perturbation theory. A better proof, based on quantum electrodynamics obeying the Garding-Wightman axioms in the Gupta-Beuler gauge, was later presented by Strocchi and Wightman [162].

Similarly, a gauge field theory of nucleons will lead to the absolute conservation of total baryon number, and this might be the reason why this is observed in Nature. There might also be one or more superselection rules operating for lepton number. An ambitious programme to derive such rules was initiated by Haag's algebraic quantum field theory. Like a group, an algebra often has many inequivalent representations on Hilbert spaces. If these representations are labelled by quantum numbers, as they are for groups, then we might be able to explain the particles we see (electron, photon, proton, neutrino ...) as the states in these different representations of an observable field. Thus, Haag is suggesting that the reason why superselection rules exist is that the algebra of observables, $\mathcal{M}$, has several inequivalent representations; this is possible if the algebra has an infinite number of dimensions. The space of all states obtained in scattering experiments is the direct sum,

$$\mathcal{H} = \oplus_q \mathcal{H}_q,$$

of states with eigenvalues q of the superselection operator, and gives a reducible representation of $\mathcal{M}$. Any observable maps each $\mathcal{H}_q$ to itself, and thus gives a representation of $\mathcal{M}$ on the subspace $\mathcal{H}_q$; call this representation π_q, so that for each $A \in \mathcal{M}$, $\pi_q(A)$ is an operator on $\mathcal{H}_q$. Haag saw that all the algebras of operators $\{\pi_q(A) : A \in \mathcal{M}\}$, for any value of q, are algebraically isomorphic to $\mathcal{M}$, but that π_q is not unitarily equivalent to $\pi_{q'}$, unless $q = q'$. Thus, for any two q, q', with $q \neq q'$, acting on two Hilbert spaces $\mathcal{H}_q$ and $\mathcal{H}_{q'}$, there is NO unitary operator U such that

$$U\pi_q(A) = \pi_{q'}(A)U \qquad\qquad \text{holds for all } A \in \mathcal{M}.$$

This is a non-trivial insight. Researchers [53, 54, 38, 37] following Haag's programme have been able to prove that if such representations exist, then particles of spin $1/2, 3/2, \ldots$ must be fermions and those with integer spin must be bosons. More, to each particle there must be an antiparticle of the same mass and spin, and the S-matrix must be invariant under PCT; this is the product of parity, charge conjugation and time-reversal.

Wightman formulated the *hypothesis of commuting superselection rules*. He noticed that in all known cases of superselection rules, the superselected operators, (charge, baryon number, fermion number, ...) commute with one another. This is the same as saying that the multiplicity of each irreducible representation in the Hilbert space is exactly one; we do not need more than one copy of each state. I should say that this is not true in a theory of non-abelian gauge fields: we need the whole non-abelian gauge group to commute with the observables in such a theory, leading to multiplicity of states labelled, for example, by *colour*. In the standard model, the quarks are confined, so this multiplicity probably does not occur, except at very high temperatures. The issue of the multiplicity of the states might be resolved by experiment, in that the Gibbs states would depend on the multiplicity.

If we adopt Wightman's hypothesis, then the direct sum of superselection sectors, with each occurring once, is exactly the simultaneous diagonalisation of the superselection operators. They are, therefore, classical variables. One can go further, and explore the idea that macroscopically distinct states of a quantum object are separated by a superselection rule.

There has been much interest in the decoherence that afflicts large quantum objects, as predicted [88]. This part of quantum theory has been called 'the theory of measurement'. The states of a classical object such as the various readings of the pointer of a measuring device, are described by commuting collective variables of the full quantum theory, taken to be the atomic system plus measuring device. The system is assumed to interact with the device in some way. The measuring device is described by a density operator on the full theory. When restricted to the collective degrees of freedom used to describe the measuring device, its density operator is assumed to become more nearly diagonal the larger the classical object is. For macroscopic bodies, the remaining coherence (that is, meaningful phase relations between its wave function and the rest of the universe) become negligible. The classical limit arises when the apparatus is taken to the limit of infinite size; then the projection operators down the diagonal can be regarded as quantum states separated by a superselection rule. A classical observable taking various values can be represented by a diagonal operator in the direct sum of the sectors. Each sector is labelled by the value of the classical variable, which commutes with all the observables. Thus, the classical variables form the *centre* of the representation of the algebra of observables.

This idea resolves the "paradox" of Schrödinger's cat. If we consider the cat as described by quantum theory with no superselection rules, then the quantum interference between the live and dead states of the cat predicts effects that are incompatible with everyone's experience of cats. But the cat is a macroscopic object; it cannot be described by a wave-function: there is no well-defined number of degrees of freedom. When a whisker falls out, bang go 10^{18} degrees of freedom, but the cat is still the cat. All we need to describe the cat in the box is a single classical variable, which is 0 if the cat is dead, and 1 if the cat is alive. The Hilbert space of the atomic system + cat is the direct sum of two copies of the system space, labelled by the set $(0, 1)$; we then impose a superselection rule between them.

The idea, introducing superselection rules between classical objects, is not apparently supported by Omnès [131] p 277. He remarks that the phenomenon of decoherence "... is also sometimes called a spontaneous generation of a superselection rule, but this calls for a notion that is not really necessary in the present case as well as an applicability which is open to controversy". If the controversy he refers to was that instigated by Bell [19], who might have been responding to Hepp's ideas on decoherence, then I disagree with Omnès. Bell argued that decoherence between classical objects is true "for all practical purposes", FAPP, but that this did not solve the matter in principle. The atomic system plus cat satisfies the Schrödinger equation, and this maintains for all time any coherent relations in the system. So the off-diagonal elements of the density matrix of the collective variables might be very very small, but a sufficiently subtle observable would reveal them. However long is the time we wait for the cat + system to settle down, as long as an infinite amount of time is not allowed, Bell can find a self-adjoint operator that has non-zero matrix element between the dead and alive cat; by measuring it, we demonstrate quantum coherence. Omnès points out p 306 that Bell "...relies on an axiom, ... namely that every observable is measurable". Omnès then shows that to distinguish energy-levels in a plausible model, so as to detect remnants of quantum phase between two macroscopic objects, would need an apparatus containing $\exp\{10^{18}\}$ particles, compared with 10 to the power of only 80 particles in the visible universe. He concludes p 319 that "the projector upon a specific energy eigenstate of the environment is an observable that cannot be measured" and concludes that this was not noticed by Bell. The problem is not with quantum mechanics, but cosmology. Omnès also mentions another contributor to decoherence, the random outside world: cosmic rays, ..., the rumble of a far-away train. It seems to me that these effects really truly destroy the phase of a classical object. This justifies the assumption that two states of a system differing in the value of a classical variable are separated by a superselection rule.

We may try to model the cat paradox theoretically, using the philosophical device of the slippery slope. This is what Bell was doing, in fact. If we succeed, we also refute the theory of phase transitions in macroscopic systems. Start by coupling the radio-active atom to a small quantum system, crudely modelling the cat and the phial of poison, and identify two states as dead and alive. We start from the undecayed state of the radio-active atom, and the cat in its alive state, and choose the best pure state we can for the rest of its attributes. Then compute the exact, pure state after one half-life. Then we increase the number of atoms in the model cat, go back to time zero, and again compute the exact state after one half-life. We repeat, increasing the number of degrees of freedom. At each stage, the model of the cat has two collective states, dead and alive, and the state after one half-life is a coherent mixture of them. We then consider the limit as the number of atoms used to describe the cat goes to infinity. Where do we put the cut between the atomic system (the decaying atom) and the measuring device (the cat)? The complex phase between the dead and live states becomes very dependent on the number of atoms, and on details that we put in at each stage about the initial state of the cat and phial. The project becomes uncomputable in the thermodynamic limit. Omnès argues that the phase is a chaotic variable, requiring more and more accuracy in specifying the initial state as the cat gets larger. After some time, but long before the cat is of realistic size, the computation will require a computer larger than the universe.

In answer to Omnès's caution in not calling this a superselection rule, I claim that it is better physics to take the limit to an infinite cat, and not to worry about the small phases at the intermediate stages. The mathematical limit to an infinite cat is an idealisation; the measurement of the cat's health uses reversible dynamics with a specific interaction at each stage of the model, and so can be said to be dynamical. However the errors in the coherence arising thereby are among the smallest in physics, one in 10 to the power of 10^{18}. The finite model of the cat used on the way is unlikely to be at all realistic, and, except that it was alive at time zero, its initial state is unknown. Nor do we know the effect of the random environment. These effects are not relevant to the conclusion, just as the shape of a calorimeter is not relevant to the measurement of the heat-capacity of an object being tested.

Bell's argument was that, although the superselection rule is there FAPP, it is not *theoretically* there. But then, the same argument could be applied in statistical physics, to show that phase transitions do not occur. To get a phase transition, we *idealise* the description, and take the thermodynamic limit to a system of infinite size. This produces some classical labels, identified by Ruelle as the algebra at infinity [142]. States with different classical values are in different thermodynamic phases. If we, with Bell, argue that in reality, the sample is not infinite, then there can be no phase transitions. In the same way, unless we idealise the measuring instrument (the cat) and model it by its essential features, $\{0,1\}$, dead or alive, then we don't get a viable theory of measurement. If we try to keep everything, we get a mess.

In [171], Wightman introduces 'new' superselection rules; here, there is something to be said for Omnès's caution. This concerns, not the decoherence of a large system, described above, but the decoherence of a mesoscopic system; he applies the idea to a single quantum molecule, which has the following strange property. It obeys quantum electrodynamics, with the conservation of parity. The ground state is invariant under the space-reflection, as can be proved for such a system. The first excited state changes sign under parity (and has no multiplicity). The difference in energy between the first and second eigen-value is so small that even in a near vacuum, the ground state is excited by absorbing say photons from the ambient electromagnetic field. It is claimed by experimentalists that it is not possible to isolate the ground state; the molecule observed is either left-handed or right-handed; each is a mixture of the first two eigen-states. Wightman suggests that when this occurs, then the left- and the right-handed versions are separated by a new superselection rule. Jona-Lasinio et al. [96] offer a dynamical reason why this occurs. However, Wightman noticed that the superselection rule has not been experimentally confirmed yet.

Penrose [133] asks for a generalisation of both classical and quantum mechanics. The new theory should reduce to quantum theory for small systems, and to classical theory for large systems. I claim that the algebraic theory, with superselection rules, does exactly this. To convince Roger is, however, a lost cause. I take comfort from 'tHooft's dictum

"The theorems and results of mathematical physics are not theorems and results about Nature. They are theorems and results about our description of it".

If I say that, in the Copenhagen interpretation, we model the measuring apparatus by classical probability, then there is no quantum interference between its readings .... by definition! We rely on Omnès, Hepp, Hartle and Gell-Mann to do the work; they show that the model, the absence of coherence, could well be the most accurate approximation of any ideal model in any branch of physics.

This division of the space of classical readings into spaces between which we have a superselection rule is imitated by the "many-worlds" interpretation of quantum mechanics; this version of Everett's "relative state" interpretation has been advocated by deWitt and Wheeler. There, it is seriously maintained that after a measurement, all the possible events actually occur, and fall into parallel realities between which there is no interaction. The main difference between this and the Copenhagen interpretation is that in the latter only one of the possible outcomes occurs, and the others do not. The many-worlds interpretation is a lost cause, since it goes counter to the Copenhagen view that the interpretation of the theory must be the same as in the classical probability theory defined by the complete commuting set: in classical theory, we do not usually try to argue that all histories in a classical stochastic process actually occur.

4.7 Measurement and Events

Harold Macmillan was asked by a journalist what, in the absence of ideology, determines British government policy; he replied "Events, dear boy, events". Some treatises on the foundations of quantum theory get their motivation from the puzzle of where events, the "quantum record", enter the Schrödinger equation; in the Copenhagen interpretation, events are created by a measurement.

Heisenberg has described the simplest quantum measurement in [86]:

"... imagine a photon which is represented by a wave-packet built up out of Maxwell waves. By reflection at a semi-transparent mirror, it is possible to decompose it into two parts, reflected and transmitted packets. There is then a definite probability of finding the photon either in one part or in the other part of the divided packet. After a sufficient time the two parts will be separated by any distance desired; now if an experiment yields the result that the photon is, say, in the reflected part ..., then the probability of finding the photon in the other part of the packet immediately becomes zero. The experiment at the position of the reflected packet thus exerts a kind of action (reduction of the wave-packet) at the distant point occupied by the transmitted packet, and one sees that this action is propagated with a velocity which is greater than that of light."

Perhaps the phrase "reduction of the wave-packet" occurs here for the first time. This passage has led to the erroneous claims of "spooky" action at a distance, and has created quite a few lost causes. Heisenberg himself did not fall into this trap, for he continues

"However it is also obvious that this kind of action can never be utilised for the transmission of signals so that it is not in conflict with the postulates of the theory of relativity".

That this conclusion is "obvious" has resulted in the example's being overlooked as a possible paradox: the collapse of the wave packet on the far side of the mirror can be known to an observer there, Bob, if he sets up detectors; if he does not, then he must wait for the observer on the reflected side, Alice, to communicate with him at a speed less than that of light. This resolution of the paradox thus involves the following ideas:

1. There are two observers A and B, each with her or his own local information.

2. The device has created two possible events; the photon will either activate a counter on front of, or behind, the mirror, but not both.
3. There is a probability (in the classical sense) of each outcome.
4. A can obtain information about the out-come by measurements, and her conditioned states need not be the same as B's, e.g. if he makes no measurement.
5. For this particular set-up, A can find out which event occurred all over by a local measurement near her. This enables her to predict what Bob will find if he make a measurement. This can be misinterpreted as spooky action-at-a-distance.

Thus we need to invoke "probability with two observers" in order to conclude that "obviously" nothing travels faster than light. It is interesting that Stapp [152], p 96, explains why a similar experiment does not require faster-than-light influences. He considers the observation of an outgoing spherical wave representing an electron. "The waveform ... suddenly collapse[s] to a small region the size of the water droplets when the corresponding track in a cloud chamber is observed. This collapse is completely natural for a probability function ... and there is no tendency ... for a quantum to be observed in one place immediately after it is observed in a faraway place." Here, he is referring to the impossibility of finding, in this example, any correlation between two separated particles forming one system, which is what happens in the EPR 'paradox'. He fails to note, however, that the single particle's presence or absence at the faraway place is 100% anticorrelated with its presence or absence at the one place. As his theory has only one observer, he should have remarked, as did Heisenberg, that the waveform in one place is changed faster than light from the faraway place could have reached it. Instead, Stapp asserts that the collapse is not only natural, but is "completely natural". He contrasts this (same page, §4.5.12) with EPR: "The nonlocal connection apparently demanded by Bell's theorem arises only after two systems originally in close communion move apart." We shall see in Chap. (6) that this paradox is also easily resolved by using two observers.

Einstein had given a thought experiment undermining reduction as early as 1927; he was worried that it changed the wave-function all over. Popper, [136], p. 235-236 knew the answer to this. The solution is "clear, if not trivial ... Saying that the logical consequences of ... this information ... spread with superluminal velocity is about as helpful as saying that twice two turns with superluminal velocity into four".

Popper has likened quantum measurement to conditioning: "the probability of an event before its occurrence is different from the probability of the same event after it has occurred" [136], page 157. Namely, the latter is 1. Even more explicitly, he wrote "reduction of the wave-packet is not an effect characteristic of quantum theory; it is an effect of probability theory in general" [137] p. 74. If he had added that, even in classical probability, there might be two observers with different information, and whose conditional probabilities are therefore different, Popper would have had a robust solution to the "problem of the collapse"; this would work when the set-up is not as obvious as in the example of Heisenberg. Then he might have been attacked by realists who would say that Popper's quantum state was subjective. All conditional probabilities are subjective, to the extent that they depend on the information (about objective facts) available to each observer. We must distinguish this from the Bayesian approach: in the Copenhagen interpretation, the observer conditions the physical state; in Bayesian theory, the information is used to condition the prior probability, which is made up by the observer.

Now come some maths! When a quantum system interacts with a macroscopic object and survives, and separates away from the macro-object, the new state of the quantum system is the marginal state of the resulting composite system. In general it is a mixed state, even if the initial state is a pure state; thus this change is not governed by the Schrödinger equation for the system, which must preserve the purity of states. A good model for this type of change uses the class of maps (on the algebra) called completely positive stochastic maps.

Definition 1. A linear map $T : \mathcal{M} \to \mathcal{M}$ is said to be *positive* if TX is positive semi-definite whenever X is positive semi-definite. A positive map $T : \mathcal{M} \to \mathcal{M}$ is said to be stochastic if $TI = I$; it is said to be completely positive if for any integer n, $T \otimes I_n$ is a positive map on $\mathcal{M} \otimes \mathcal{M}^n$.

In one time step, the interaction of an atomic system with a heat bath can be described by some stochastic map T, and the sequence of maps

$$A \mapsto TA \mapsto T^2 A \mapsto T^3 A \ldots$$

is an example of a discrete-time quantum stochastic process. The operator $T^j A$ can be taken as the time evolution of the operator A in j time-steps, in the presence of a noisy environment. The usual unitary evolution in the Heisenberg picture,

$$A \mapsto U^{-1}AU \mapsto U^{-2}AU^2 \mapsto U^{-3}AU^3 \ldots$$

is an example of this, when there is no noise. Another example of a stochastic map is any convex mixture of Heisenberg maps

$$TA = \sum_i \lambda_i U_i^{-1} A U_i, \qquad \text{where } \sum_i \lambda_i = 1 \text{ and } \lambda_i > 0.$$

In this way, we can regard dissipative dynamics as random Heisenberg dynamics, by interpreting λ_i as the probability that the dynamics is given by U_i. In this example, T is a stochastic map, which is not only positive but is completely positive. This is needed; for $T \otimes I_n$ is the action of T, extended by the identity I_n on any other system not mentioned in the algebra $\mathcal{M}$ on which T acts. This combined action should be positive too, for any n. Note that when

T is not invertible, which is when dissipation occurs, then it is not a unitary conjugation, and so is outside Schrödinger's formulation of time-evolution.

A stochastic process also can be expressed in terms of the dynamics of the states, using duality. Thus, the dual action, T^*, on the states is the map $\rho \mapsto T^*\rho$, where $T^*\rho$ is the map $A \mapsto \rho(TA)$. In fact, the object $T^*\rho$ is indeed a state. For

$$T^*\rho(I) = \rho(TI) = \rho(I) = 1,$$

since $T(I) = I$. More, if $A \geq 0$, then

$$T^*\rho(A) = \rho(TA) \geq 0$$

since $T(A)$ is positive.

For example, for unitary maps, for which $TA = U^{-1}AU$ we get that $T^*\rho A = \rho(U^{-1}AU)$, showing that

$$\mathrm{Tr}\,(\Delta_{T^*\rho}A) = T^*\rho(A) = \rho(TA) = \rho(U^{-1}AU) = \mathrm{Tr}\,(\Delta_\rho U^{-1}AU)$$
$$= \mathrm{Tr}\,(U\Delta_\rho U^{-1}A)$$

for all A. Since a state is defined by its action on all observables, we get (from the last term) von Neumann's formula for the evolution of the density matrix $\Delta \mapsto U\Delta U^{-1}$ in one time-step.

In the Copenhagen interpretation of quantum mechanics, a measuring instrument is not just any old macroscopic system with which the quantum system interacts. It is a cleverly designed instrument of macroscopic size, which responds sensitively to the state of the atomic system. A heat bath is the antithesis of a measuring instrument; being infinitely large, it is unchanged by the atomic system. An instrument must be designed for the purpose of measuring a particular observable, X say. We should consider a sequence of instruments of increasing size, which defines the ideal instrument only in the limit to infinite size. We have argued above that according to Hepp [89], the system plus finite measuring device can be treated as a quantum system, obeying von Neumann's equation, which is linear in the state and is given by a unitary group. The measuring device should have a collective variable, call it Z, which remains there whatever the size of the instrument and whose values become labels for the positions of the pointer of the dial in the infinite limit. For all finite sizes, the coupling between the quantum system and the instrument are such that after a short interaction time, there is a strong correlation between the eigenstates of X and the corresponding values taken by the collective variable. We can estimate the coherence between the different dial readings, for each finite model instrument, by the size of the off-diagonal entries in the density operator, in a basis in which both X and the dial variable Z are diagonal. In Hepp's model this coherence converges to zero as the size of the instrument gets larger. The infinite limit of the model, restricted just to the algebra generated by X and Z, shows the following features, which I call *ideal*.

When the system in the state ρ is passed through the instrument, and the instrument is observed, the dial is found to show one of the possible values $\Omega := \{x_1, x_2, \ldots x_n\}$ of X. (For simplicity, I have assumed that none of the eigenvalues is multiple). The state of the emerging system also changes, to the corresponding eigenstate of X. According to the Copenhagen interpretation, the probability of this is determined by ρ. To find this classical probability, use ρ to determine the expectations of the powers of X. The sequence $\{\rho(X^j)\}_{j=0,1,\ldots}$ is the moment sequence of a random variable (also called X) on the sample space, the spectrum of X. This result follows from the positivity of ρ on any positive operator of the form $Z = Y^*Y$, and $Y = \sum \alpha_j X^j$ for some complex numbers α_j. The probability measure, that the dial will show the reading corresponding to x_i, is then that determined by this moment sequence.

To clarify the measurement process, we divide it into two steps. In the first, the state ρ is passed through the apparatus designed to measure A, but the dial is not read. In the second stage of the measurement, the dial is read. This recalls a suggestion of von Neumann, that the dial of the instrument should be read by a computer; this was counter accusations of subjectivity, a fate that I do not mind.

In the first stage, the state of the system is changed by a completely positive stochastic map; according to von Neumann, T^* is given by $T^*\rho := \sum_i P_i \rho P_i$, where P_i are the spectral projections of X, the observable that the instrument is designed to measure. Note that $T^*\rho$ is linear in ρ. Thus, models of measurement that try to describe the apparatus by a nonlinear operator are lost causes. By the assumption made here, each projector P_i is one-dimensional, projecting onto ψ_i. However, the measurement process is not merely the map $\rho \mapsto T^*\rho$. The latter correctly represents ρ as the statistical mixture of the possible outcomes, eigenstates ψ_i, of X, with the correct weight. But, there is nothing about the state $T^*\rho$ to tell us that it these states, ψ_i, rather than those of a different decomposition of ρ, that will be found. It is the dial of the measuring apparatus that tells us which decomposition is realised. There is a 100% correlation between the classical dial reading x_i and the component $P_i \rho P_i$ of $T^*\rho$ that emerges as the quantum state when x_i is observed.

In the ideal measuring instrument, the details, e.g. the quantum nature, of the classical apparatus do not matter; this indifference is comparable to that arising in classical physics. For example, we get the same answer for the heat capacity of an object whatever the shape of the calorimeter. The Copenhagen school does NOT attempt to give more details of the apparatus than that it is able to show the values x_i, correlated with the corresponding eigenvector. The Hilbert space $\mathcal{H}$ of the system is the span of the ψ_i, and the state-space of the system plus apparatus is the direct sum $\mathcal{K}$ of n copies of $\mathcal{H}$, each one labelled by one of the possible configurations x_i of the dial:

$$\mathcal{K} := \mathcal{H} \oplus \mathcal{H} \ldots \oplus \mathcal{H} \ (n \text{ copies}).$$

Since the dial is a macroscopic object, there is a superselection rule preventing any quantum interference between these components of the direct sum. Moreover, the component of the state seen in the ith copy of $\mathcal{H}$ must be $P_i \rho P_i$. The apparatus plus system is described therefore by the C^*-algebra $L(\Omega) \times \mathcal{M}^n$, and the state on it is the density matrix

$$\sum_i \delta(x - x_i) P_i \rho P_i.$$

A similar conclusion has been found by Araki [5, 6]; he constructs a model with continuous values for a superselection rule. His technique is to take the limit, to infinite time, of the dynamics of a spin system measured by a classical magnetic field.

The measuring apparatus designed to measure X, using the abelian algebra of compatible observables, has been used in Sect. (3.6) to create Gelfand's sample space $\Omega = \{x_1, \ldots, x_n\}$, where each x_i is a character. Thus *events* can be identified. These events are the subsets of Ω, so exist in the correct mathematical sense, while without measurement, the quantum theory had no events. Note that events are not created by an observer's *thinking about* doing a measurement, and mentally writing ρ as one rather than another decomposition into pure states. It is the switched on apparatus that changes ρ and creates events.

The events occur in the physical sense when we pass to the second stage of the measurement, the looking at the dial. The probability that x_i is shown on the dial is $p_i = \text{Tr}\, P_i \rho P_i = \text{Tr}\, \rho P_i$. We then obtain information within the entirely classical theory (Ω, p), and this information causes the probability to be conditioned by Bayes's formula. By looking at the dial and finding x_i we know that the system is in the state ψ_i. Our assignment of density operator changes from Δ_ρ to P_i. This is the collapse of the quantum state; it is just Bayes's formula for conditioning in the classical theory created by the choice of complete commuting set. If quantum theory has any puzzle, this is also a puzzle for classical probability. The only difference is that, in quantum probability, there is more than one complete commuting set, and each gives a different sample space and probability: the statistical model is *contextual*.

If we obtain only partial information, that the dial reading was in some subset E of Ω, then we would say that the event E had happened; we can note this down and it becomes part of the *quantum record*. Partial information like this causes a partial collapse of the state, from ρ to

$$\frac{\sum_{i:x_i \in E} P_i \rho P_i}{\sum_{x_i \in E} \text{Tr}\, P_i \rho P_i}.$$

This is the quantum Bayes conditioning formula, offered by Lüders [112] as a natural generalisation of von Neumann's formula. That this is the quantum analogue to Bayes conditioning has sometimes been attributed Bub [36], though it was known to von Neumann to reduce to Bayes conditioning in the commutative case. If E contains only one element, x_i, this reduces to P_i.

In this account, we assumed that X had simple spectrum. This is equivalent to assuming that $\mathcal{P}(X)$, the polynomial algebra generated by X, is maximal abelian (is not contained in a larger abelian algebra). In Dirac's terms, we assumed that X was a complete set of compatible observables. If the spectrum of X is not simple, we can add enough compatible observables so that we get a complete set. Then they can be simultaneously measured by a clever piece of apparatus. This will have several dials, one for each element of the maximal abelian set. A similar argument shows that if we look at some of these dials but not the rest, then we observe events and can find the conditioned quantum state. A superselection rule then arises between states with a different value of any one of the measured observables.

However, if the apparatus measures only X, which has some multiple eigenvalues, and does not measure the other observables making up the complete set, then the quantum phases are only destroyed between wave-functions associated with different eigenvalues of X; these events are separated by a superselection rule. The eigenspaces of X remain coherent, and the classical event that has happened does not give a complete picture of the collapsed state, which is not always pure if the eigenvalue observed is multiple.

Omnès and many others conclude that the collapse of the wave packet is not "physical". We do not agree. In classical probability, conditioning implements the change in information acquired by the observer, so this must be the case also in quantum probability. But there, unlike the classical case, the effect of the measuring device, T, is physical: it changes the prior state, from ρ to $T^*\rho$.

We can understand why the idea of collapse arose. This is seen when the device is more like a filter than a dial. Suppose the observable being measured is P, a projector, and suppose the observer sorts the events showing value 0 for P from the events showing 1, and removes those showing 0 from the flow of samples. Then a weaker beam of particles would be best represented by the state given by $P\rho P/\mathrm{Tr}\,[P\rho P]$, rather than $P\rho P + (1 - P)\rho(1 - P)$. This selection can be done entirely automatically, as in a Nicol prism, which absorbs photons with a certain plane of polarisation, and transmits those with the orthogonal polarisation. It is clear then that the measurement has had a physical effect, and that this is given correctly by the collapse postulate. The example of the Nicol prism was used in the first days of quantum theory, [52] and explains the physics behind the collapse postulate.

4.8 The Quantum Game with Two Observers

Books on quantum theory nearly always assume that there is only one observer, who is able to measure any complete set of observables he chooses. Some of the "paradoxes" of quantum theory arise when the formalism suitable for the case of one observer is applied to a case in which there are two or more observers. In this section we sketch the theory of the quantum game with two observers, A = Alice and B = Bob. We assume that A can measure any observable in the algebra $\mathcal{M}_A$ and that B can measure any observable in the algebra $\mathcal{M}_B$. These will be assumed to be finite-dimensional, and are the analogues of the information sets of the two players that arise in classical game theory. There, the information set of a player can be regarded as a partition of Ω into a coarse-grained version into r parts: $\Omega = \sqcup_j E_j$. Here, the symbol $\sqcup$ denotes that the sets $E_1, \ldots, E_r$ are disjoint, and that Ω is the union of these. Only the sets $E_j, j \in \{1, 2, \ldots, r\}$, are events whose occurrence can be confirmed by that player. Any partition of Ω, say $\sqcup_j E_j$, generates a Boolean ring, say $\mathcal{B}$; it consists of all unions of sets E_i, including the empty union. The algebra of all $\mathcal{B}$-measurable functions is exactly the function space spanned by the indicator functions of the sets in the partition. In this way, in classical game theory we associate an algebra of random variables with each player. In quantum game theory, we associate a non-abelian algebra of observables with each player. If the information algebras of the two observers, A and B, do not commute with each other, then the consequences of their making observations depend on the order in which they make the measurements. We shall not discuss this case, but shall assume here that the non-abelian algebras commute with each other, contain only the identity in common, and, more, that they together generate the algebra of the system. This is true of the set up in the Einstein-Podolski-Rosen 'paradox', in the form suggested by Bohm; there, $\mathcal{M}_A = \mathcal{M}^2 \otimes I_2$ and $\mathcal{M}_B = I_2 \otimes \mathcal{M}^2$, and the full algebra is isomorphic to $\mathcal{M}^4$, the algebra of the spin observables of two spin-1/2 particles.

The game starts with the state of the system, ρ, which we assume known to both players. Let $\Delta := \Delta_\rho$ be the corresponding density matrix. Given a sample of the state ρ, Alice can measure any of her observables X, but she would not be able to measure any of Bob's observables except the sure elements (multiples of the identity). Thus, Alice is sampling the state ρ restricted to $\mathcal{M}_A$, which we shall denote by $\rho|_A$. Similarly, Bob can sample only $\rho|_B$. These define, respectively, density matrices $\Delta|_A$ and $\Delta|_B$ in the algebras of Alice and Bob.

We now consider what happens when Alice and Bob measure observables X and Y. It is possible to do this in such a way that neither disturbs the measurement of the other, since X and Y commute, and could be measured simultaneously by a clever-enough single observer. Indeed, if $X = \sum_i x_i P_i$ and $Y = \sum_j y_j Q_j$ are the spectral decompositions of commuting matrices X, Y, then by measuring X we change any state ρ to $T_X \rho$ whose density matrix is $\sum_i P_i \Delta P_i$, and if (later) we measure Y we change this state to $T_Y T_X \rho$ whose density matrix is $\sum_j Q_j \sum_i P_i \Delta P_i Q_j$. Now, the spectral projections P_i commute with the spectral projections Q_j, so we can alter the order of the products, to see that $T_Y T_X = T_X T_Y$. So the order of measurement does not matter. More, if we measure X we do not disturb the distribution of Y. To see this, let $j > 0$ be an integer. Then the mean of Y^j in the state $T_X \rho$ is

$$(T_X \rho)(Y^j) = \mathrm{Tr} \left(\sum_i P_i \Delta_\rho P_i Y^j \right)$$

$$= \mathrm{Tr} \left(\sum_i P_i \Delta_\rho Y^j P_i \right)$$

$$= \mathrm{Tr} \left(\sum_i \Delta_\rho Y^j P_i \right) = \mathrm{Tr} \left(\Delta_\rho Y^j \right)$$

$$= \rho(Y^j)$$

since the sum of the projectors is unity. Since all the moments of Y are the same in the state $T_X \rho$ as in the state ρ, the distributions are too. Thus our completely positive stochastic map T_X has the property of not disturbing the measurement of Y, as desired. If all Bob's observables commute with Alice's, then the same calculation shows that $(T_X \rho)|_B = \rho|_B$: Bob's partial state is not changed: there is no disturbance of B's conceivable experiments by the instrumentation set up by A.

After the measurement, but before the looking, Alice assigns the state $T_X \rho$ to the system, which as we have seen, has the same partial state on Bob's algebra as before the measurement. Similarly, Bob assigns the state $T_B \rho$ after he measures Y, and this has the same partial state on Alice's algebra as before the measurement. These are different states, but they genuinely (and experimentally) reflect the correct assessment of A and B to the chances of various dial readings. Contrast this set-up with the case when there is only one observer, measuring both X and Y. He has a single state, on the whole algebra. This state is not the same as that of either A or B, which are states on subalgebras. This state of affairs is often met in the theory of quantum information channels. Some of the trouble in papers on the quantum paradoxes arises because the authors assume that all three states are equal, which is tantamount to assuming all observers use the same channels.

Suppose now that X and Y (lying in $\mathcal{M}^2$) have simple spectrum, the two eigen-values (x_1, x_2) and (y_1, y_2). To describe the statistics, Alice must set up the direct sum of 2 copies of the Hibert space of the system, $\mathbf{C}^4$, with a superselection rule between them, coming from the fact that each copy corresponds to a different value for her macroscopic variable, the dial reading, x. She does not set up superselection rules for Bob's dial readings, since she does not know what measurement he has made. Alice must enlarge her algebra to include the classical variable, x, which is the diagonal matrix $\Xi := \mathrm{diag}\,[x_1, x_2]$, and all its polynomials, which form the classical algebra $\mathcal{P}(X)$. The algebra is thus temporarily enlarged to $\mathcal{P}(\mathcal{X}) \otimes \mathcal{M}^4$. Because her experimenter is very skilled, the dial reads x_1 only when the value of X is x_1. Thus not all states of the enlarged algebra are observed.

When Alice looks at the dial, she reveals the value of a classical object, the dial, to herself and those close to her. She does not have to reveal it to Bob, and, according to special relativity, she cannot do this instantly if Bob is far away. Bob himself can measure any Y that he chooses, and obtain a new, updated state to replace $\rho|_B$. This does not affect Alice's state, whether he does it before, or after, her own measurement, or not at all. In the *EPR* experiment, there is a strong correlation in the state ρ between the observables X and Y. In the next chapter we shall see that A can obtain information about B's measurement, if he were to do one, but she has no influence on what B decides to measure, or on its result.

The idea of attaching a state to an observer rather than to the system, we solve the paradox of Wigner's friend; in this paradox, Wigner sets up the measuring apparatus and pushes the start-button, but his friend reads the pointer. Does the wave-function collapse or not? The answer is that Wigner assigns the state $\sum_i P_i \rho P_i$ while his friend assigns $P_i \rho P_i / \mathrm{Tr}\, P_i \rho P_i$ where i is the result seen by Wigner's friend.

Again, we can imagine circumstances in which the collapse of Alice's state to the measured eigenstate is physical, but Bob's uncollapsed state is also physical. This uses the *EPR* experiment again. Suppose that an atom at the origin decays into two photons, that move off in opposite directions to far-away experimenters, Alice and Bob. Suppose that there is a 100% correlation between the spins of A's and B's particles, and that Alice measures her spin using a Nicol prism, and B makes no measurement. Alice will get the state $P\rho|_A P + (1-P)\rho_A(1-P)$, but the filter allows only the state P through; the other spin is absorbed by the filter, and she will renormalise her state to get the reduced state $\rho_{\mathrm{red}} := P\rho_A P / Tr(P\rho|_A P)$, while B will use the state $\rho|_B$ to describe his continuing beam with both sorts of spin. An observer C watching the combined system describes the state by the tensor product $\rho_{\mathrm{red}} \otimes \rho|_B$, showing that the state has reduced for Alice but not for Bob.

5

Bayesians

The idea of Bayesian statistics is related to that of Laplace, and in fact pre-dated it; it claims that if we know nothing about the die we are casting, then the best chance is to assume that all the numbers, $\{1, 2, \ldots, 6\}$ are equally likely. Modern Bayesians generalise this idea to infinite dimensions, and even to systems with continuous random variables. I claim that this generalisation is not permitted in general.

5.1 Classical Bayesianism

Consider the following paradox, told to me by Dr. A. Barnard. A player, A, selects a positive number x at random, and then writes out two cheques. One cheque is for the number, £x, and the other is for double this, £$2x$. He puts the cheques in two envelopes, and invites his opponent, B to choose one, open it and see how much the cheque is for. Say it is for £y. A does not tell his opponent, B the value x that he first thought of, but invites him to choose, either to accept the cheque he has to hand, or to burn it, and choose the other cheque. This will be, either £$2y$ or for £$\frac{1}{2}y$, each with probability $\frac{1}{2}$. One sees that the average gain to B if he burns the first cheque he gets, and chooses the other envelope, is

$$\frac{1}{2} \cdot 2y + \frac{1}{2}\frac{y}{2} = \frac{5}{4}y > y.$$

The average gain if he keeps the original choice is y, which is less. However, there is complete symmetry between the two choices, and so the gain of switching envelopes should be y. Where does the asymmetry arise?

We see that B has to compute the probability that the unseen envelope contains $2y$, and $y/2$, given that the partner envelope contains y. He knows that he sees either $y = x$ or $y = 2x$. If these are of equal probability, then indeed the probabilities that the other one contains $2y$ is the same as the probability that it contains $y/2$. Indeed, if the probability that A chose x is $p(x)$, then the probability that B chose the envelope containing x is $\frac{1}{2}p(x)$

and the probability that he chose the envelope containing $2x$ is $\frac{1}{2}p(x)$. It is a mistake to assume that these are the same. In the first case, $y = x$ and in the second case, $y = 2x$. B knows only y, not x. Thus the probability of the first case is $\frac{1}{2}p(y)$ and the probability of the second case is $\frac{1}{2}p(y/2)$. These are only equal if $p(y) = p(y/2)$. To work for all cases of the game, we need $p(x)$ to be independent of x. Thus, this hidden assumption was implied by the argument given above. In fact, it is not possible to assign the uniform probability to the choice of x by A in the first place: we cannot have $\int_0^\infty p(x)\,dx = 1$ for any constant function $p(x)$. We could discretize the problem; in order for both $\frac{1}{2}y$ and $2y$ to be possible values, given that y is a possible value, the space Ω must contain the set

$$\Omega = \left\{ \ldots, \frac{1}{8}, \frac{1}{4}, \frac{1}{2}, 1, 2, 4, 8, \ldots \right\}.$$

It is not possible to assign the uniform probability on this set, because it contains infinitely many points; we need a finite sample space to do this. However, any finite space leads to an asymmetry between the first and the second choice of B, given that he reads the amount on the cheque. Suppose that

$$\Omega = \{1, 2, 4, \ldots, 2^{N-1}\}.$$

We impose that A choose randomly from this set with uniform probability, $p = 1/N$, and writes two cheques, for his choice, x, and for $2x$, and poses the same problem to B as above, except that the space Ω and the probability p on it are now made explicit to B. A changes the rules of the game for at least one choice of x: suppose if A choose $x = 2^{N-1}$, the largest possible, then he writes the second choice for half of this, 2^{N-2}, instead of double. If the first choice of B is for the envelope containing $x < 2^{N-1}$, we write this as $(x, 2x)$, while if B chose the envelope containing $2x$, we write this as $(2x, x)$. Suppose that by chance, equal to $1/N$, A chose $x = 1$. Then B receives two envelopes, and he has the choice, each with the probability $1/2$, of making the choices $(1, 2)$ or $(2, 1)$. If B makes the choice $(1, 2)$, he sees the value $y = 1$ and as he knows that Ω does not contain the element $\frac{1}{2}$, he knows that he must burn the cheque for £1, and switch his choice of envelope to do better. If he by chance opened the envelope containing 2, so that he chose $(2, 1)$, then he knows that the other envelope contains 1 with chance $\frac{1}{2}$, but might contain £4 with the same chance; so again he burns the first envelope, and increases his mean winnings to

$$\frac{1}{2} \cdot \frac{1}{2} + \frac{1}{2} \cdot 2 = \frac{5}{4}$$

pounds. B does the switch, gaining a better mean, all the way up to seeing 2^{N-2}. But finally, if he finds that $y = 2^{N-1}$, he knows that to switch will lead to the number 2^{N-2}, a very much smaller number, and he does not make the switch in this case. This difference spoils the symmetry between the two choices that B can make, and removes the paradox. It is no use taking N to increase to ∞: then $1/N \to 0$ and we are not left with any model.

The model of the paradox, that all $x > 0$ are equally likely, could be changed. For example, it might be thought that the larger the value of x the smaller should be its probability. Jeffreys [95] proposed that a random variable, taking positive values, but otherwise completely unknown, should have the probability density proportional to $1/x$. This is not normalisable. It turns out that when this density is used in the above game, instead of the uniform density for x, then the paradox does not arise: the expectation of the winnings of B are the same whether he changes his envelope or not. However, this does not save the idea; it is a lost cause, because $1/x$ is not the density of a normalisable probability.

A similar idea arises when we consider x to be the probable life-time of an object, about which we know nothing. In a recent article, [102] Kierland and Monton claim that the correct Baysian probability for the unknown life-time remaining for an object, given its present age, should be larger the larger the given present age is (no other facts being known). This idea is illustrated by considering geysers found in an unknown desert in an unknown country (such as New Zealand). If one is told by the tourist guide that of two otherwise similar geysers, one was known to have started up in its present state ten years ago, whereas another sprang into existence ten minutes ago, then one should think that the older one is more likely than the new arrival of still being in existence in say 20 years' time. The article gives a formula, due to Gott [74] for the probability distribution of its lifetime, given the present age, but assuming no other knowledge at all about geysers.

Gott argued as follows. Assume that we observe the phenomenon in question (above, the activity of the geyser) at a time t between t_0, when it began, and t_1, when it ends. The principle of Copernicus then says that in the absence of evidence to the contrary, we should not think of ourselves as having a special position in the universe. Thus, t should be uniformly likely between t_0 and t_1: it has the uniform distribution. Thus,

$$r := \frac{t - t_0}{t_1 - t_0}$$

is a random variable with the uniform distribution on $[0, 1]$. For example we assign 95% probability to the proposition

$$0.025 \le r \le 0.975.$$

Put $t_p := t - t_0$, the time past, and $t_f := t_1 - t$, the future time available. Then Gott's assumptions lead us to the equation

$$\text{Prob}\left\{0.025 \le \frac{t_p}{t_p + t_f} \le 0.975\right\} = 0.95\%.$$

Take the inverses, to get

$$\text{Prob}\left\{\frac{40}{39} \le \frac{t_p + t_f}{t_p} \le 40\right\} = 95\%,$$

which says that with 95% confidence we have $\frac{40}{39}t_p \leq t_p + t_f \leq 40t_p$; that is

$$\frac{1}{39}t_p \leq t_f \leq 39t_p \tag{5.1}$$

holds with 95% confidence. The choice of 95% confidence is not essential: one can similarly find the confidence interval corresponding to any other choice.

Bayesians interpret Eq. (5.1) as a probability for values of t_1, given t_0 and t. Caves [41] says that this "has no justification in probability theory". Kierland and Monton retort with "that's because it is justified in arithmetic". This shows the difference between the rest of us and Bayesians. For most of us, t_0 and t_1 are parameters, between which the random variable t lies (with uniform distribution, in Gott's model). We use the one piece of data, the present age, to find the interval that remains of the life, consistent at the 95% level. However, Bayesians assume that t_0 and t_1 are random variables with some distribution, or at least, can be treated as such. It turns out that Gott's assumption about the distribution of t requires that t_0 must be uniformly distributed and that the lifetime, $t_1 - t_0 = t_p + t_f$ should have the Jeffreys density; neither is normalisable, so strictly speaking, the subject is outside Kolmogorovian probability. More work is needed. Kierland and Morton remark that such densities might be covered by the axioms of Renyi. It is clear that the meat of Gott's argument is that t is uniform; this is Gott's prior distribution in the above work.

Kierland and Monton apply Gott's formula to the problem of the geysers: for the one known to be 10 minutes old, they get for the remaining lifetime

$$\frac{10}{39} \text{ minutes} \leq t_f \leq 390 \text{ minutes}, \tag{5.2}$$

while for the older one they get

$$\frac{10}{39} \text{ years} \leq t_f \leq 390 \text{ years}, \tag{5.3}$$

each with 95% confidence. They go on to note that the formula has been applied by Gott to human affairs such as the existence of intelligent life on earth, now say 200,000 years old. The formula then says that intelligent life will (with 95% confidence) survive t_f more years, where

$$5,128 \text{ years} \leq t_f \leq 7,800,000 \text{ years}.$$

They point out that Gott realised that, if more is known about any question than the present age, then this should be built into the prior; the above theory applies only when nothing else is known except the present age.

Gott's prior can be deduced from the Jeffreys prior for the total life; such a possibility was roughly shown by Gott [75]. Jaynes has argued that the Jeffreys prior is the only density such that the predictions are invariant under transformations of location and scale [94]. For example, we saw that the predicted life-time of a geyser was independent of whether our trip to the desert was on Sunday or Monday; this shows that it was independent of location (here, in time). Also, the predicted lifetime is scale invariant, in that the ratio of maximum to the minimum times in the inequality Eq. (5.2) is the same as the same ratio in Eq. (5.3).

Gott's principle, applied to a specific radioactive substance, is false. From a large collection of radioactive atoms of the same half-life there are always some particles that have not yet decayed. However, their future lifetime is the same however long they have so far survived. The geyser principle, that a newly created atom has a shorter lifetime than one that has lasted 200 years, given both cases, is false. Another failure of the principle has been remarked by Bass [16]. He notes that it is false when applied to the question of the time of death of an animal, and how it relies on the present age. It fails to work in any question about which we have any information whatever. In science, we aim to find out much more detail than we do in, say, tourism.

I am reminded of factor analysis, a technique discovered by statisticians working in psychology. In this subject there was very little known about the *why*. In factor analysis, the correlation matrix between say 12 random variables is diagonalised and reduced to its main *factors*. These are the linear combinations of the original random variables that correspond to the few largest eigenvalues of the correlation matrix. The method was suggested as a technique for finding unknown laws of physics. For example, consider various physical objects, and measure their weight, length, breadth and depth. We find the correlation matrix, which can be diagonalised. With luck, we find it has two common factors, called "size" and "density". This result might lead to the discovery that space exists, and have three dimensions. However, science is much more accurate and detailed than such methods can reveal.

5.2 Physics from Fisher Information

An attempt to apply Bayesian information theory to physics was made by Roy Frieden in his book, Physics from Fisher Information, published by Cambridge University Press, $80. In it, he claims to derive the laws of physics from the principle of stationary information. However, these claims do not stand up to examination. Let us look at two such claims; in the first, he claims to show the foundational idea that in any measurement, the "information deficit", $K = I - J$, is stationary. Here, I is the information obtained from the measurement, and J is the "bound information", which is not exactly defined, but is something like the information that cannot be extracted by the measurement. The author tentatively identifies I with the kinetic energy, and J with the potential energy. He then says, without any comment, that $I = J$. This seems to be related to a theorem on simple harmonic motion, which says that in classical dynamics, the time-means of I and J, over a single period, are equal. Of course, there is no reason why $I = J$ should be true for other than simple harmonic motion, and even for this case, the equality holds only as a mean, and not at a general time. The author then shows that $\delta K = 0$ in two pages of Fourier analysis. However, if $I = J$, then $K = 0$, and so $\delta K = 0$ is trivial. I understand that the second edition of the book will give this shorter proof, with thanks to me; this does not mean that I endorse his claim; he assumed the answer in his assumption that $I = J$.

Another claim made in the book is that if K is stationary we are led to quantum mechanics. In fact he postulates the formalism of quantum mechanics, in particular the identification of the mean energy W in a non-normalised state ψ as the scalar product

$$W := \frac{\langle \psi, (T + V)\psi \rangle}{\langle \psi, \psi \rangle}.$$

Here, T is the kinetic energy and V is the potential energy. He then finds the turning-points as we vary ψ of $\langle \psi, (T + V)\psi \rangle - \langle \psi, \psi \rangle W$. Of course, this is zero for all ψ, by the above equation. The functional derivative of this relative to ψ is therefore zero for all ψ. In calculating it, he erroneously forgets that W depends on ψ. So what he is really doing is finding which states make the quantum mechanical mean energy $\langle \psi, (T + V)\psi \rangle$ stationary, subject to $\langle \psi, \psi \rangle =$ a constant. Then W appears as the Lagrange multiplier. His solution is that ψ must be an eigenstate of the Schrödinger operator, and W must be an eigenvalue. He then claims to have derived quantum mechanics from information theory (forgetting that he started with the Schrödinger time-independent equation). One can more soberly argue that he has proved that every measured state is an eigenstate of energy. The trouble with this idea is that the form chosen for the Fisher information, which was $I = T$, is argued earlier in the book to be that which one gets when measuring the *position* of the particle, not its energy. Measuring the position of a particle does not send it to an eigenstate of energy. As it is, his result is hardly new, being the basis of the Rayleigh-Ritz method of finding the eigen-values, known since the nineteenth century. The author also fails to mention that in the rest of the book it is $K = I - J = T - V$, the information deficit, that is supposed to be a minimum and therefore stationary, not $T + V$; but what is a minus sign between friends?

For more, see our review: D. A. Lavis and R. F. Streater, Physics from Fisher Information, *Studies in the History and Philosophy of Modern Physics*, **33, 327–343, 2002**.

5.3 Quantum Bayesianism

The first quantum Bayesian was von Neumann. In **Die mathematischer Grundlagen der Quantenmechanik** [166], he describes the measurement process of say the spin polarization of an electron source as follows. First, we know nothing, so we take the state to be the state of maximal entropy, the fully unpolarised state. In finite dimensions, this is proportional to the identity matrix. We then pass the beam from the source through a filter, P, getting the state $P\rho P/\mathrm{Tr}\,[P\rho P]$; then we refine the measurement by passing it through another filter defined by Q, compatible with P, and so on until we find a one-dimensional answer. von Neumann's idea was attacked in a brilliant polemic in [109]. Krylov complained that this method starts with the assumption that every unknown electron source is unpolarised. He says that, in fact, completely unpolarised electron sources are very rare. Of course, when we do not know the direction of the polarisation, or the extent of the polarisation, then this *physical* information becomes mixed up with the classical lack of knowledge about the orientation of the source. Krylov did not believe that all the characteristics of the state reflect only the lack of knowledge of the observer, but that there was a physical state, ρ_0 out there to be found. He said that the job of physics is to find the state at a later time, given the state at an earlier time. Bayesians, on the other hand, attribute *all* the entropy in a state to the lack of information in the observer. They start with a prior state, which incorporates all the information available to the observer, and perform measurements, and condition the density operator using the quantum Bayes rule; they can use non-commuting operators, provided that they start again with a fresh sample from the source. They then at each stage can choose a strategy to maximise the gain in information (on average) gained by the experiment, and at each state announce the most likely state given the results so far.

This looks very sensible. Except. We read that a true Bayesian should wind in to his prior all his beliefs about the state of the system, as well as the chance he gives that the theory being used is correct or not correct. It might be that there is some chance that the theory is incorrect; taking this into account will (perhaps) change for the better his assessment of the chance that a certain result will be found or not found. But physics is not a gambling game. We are not here to predict results, but to test theories. Suppose that we want to test whether a popular theory should be rejected on experimental grounds. We assume the theory is true (= hypothesis H) and find the probabilities of results $x_1, x_2, \ldots$ for an observable X, given H. Suppose that we agree to the following test: we reject H if the measurement of X gives x_1 which, under H, is less than 1% probable. If the test does give x_1, we publish this, refute the popular theory, and get famous. Suppose, however, we personally believed H only 50%, and so gave it only half a chance, as we also did a rival theory H' that allowed $X = x_1$ with 10% chance. We construct a hedged hypothesis, which assumes H with probability 1/2 and H' also with probability 1/2. Then the hedged theory would predict that the chance of x_1 is just over 5%, not significant, and we come to no conclusion. This illustrates the use of a model probability space (Ω, p) (under the hypothesis H) where $p(\omega)$ is not related to the measure of belief by the statistician in the likelihood that ω will occur. Of course, measure of belief is not a mathematical concept, and cannot be used as the definition of a probability. The reverse is the case: once a probability model $(\mathcal{M}, \rho)$ has passed some statistical tests and has been accepted by a community, then the members of that community can rationally use the predicted probabilities of the model as their measure of belief in the various events.

Krylov believed that the statement that a source, aligned in a certain way, emits polarised electrons along its direction of alignment, is a physical statement; but to say that the same source, of unknown alignment, emits unpolarised electrons, is incorrect. Here, we can distinguish between physics and calibration practice, just as J. A. Wheeler distinguished between physics and the *things of physics*. Wheeler thinks that a graduate Ph. D. in physics at Princeton should have some familiarity with the things of physics, such as how to set up a good vacuum, or to calibrate a gamma ray source in curies. Thus, Krylov thought that setting up an aligned source was a problem of calibration, to be implemented by the technician; it is one of the things of physics without being physics.

There can indeed be a difference between an unpolarised electron source and a source of a fully polarised electron whose alignment we do not know. This arises when the source is macroscopic, such as a crystal of radioactive atoms that are all aligned in the same way in the crystal. The macroscopic size of the sample requires that the mixture of all the differently oriented directions is incoherent: a superselection rule holds between the different polarisations. The orientation is a classical variable, which can be found without disturbing the system. This exercise might need theory, and a brilliant technician, but not quantum theory.

The difference between physicists and Bayesians is that physicists use ρ as their prior, whereas Bayesians use their best guess at ρ. Physicists might not know ρ; in that case, they insert parameters to describe it; or they might make a hypothesis H about ρ and apply the usual significance tests on H. Thus, there is an objective part to the physicists' ρ; it is then conditioned by a measurement. On the contrary, the Baysian's ρ is entirely about his knowledge.

The Bayesian approach to probability is often called prediction theory. It is better called hedging theory: we try to form a strategy that keeps the penalty for being wrong to a minimum. It has its uses; in physics, when thinking about building a huge accelerator to find say, the Higgs boson, we have to assess the chances of not finding it. This must mean assigning a probability distribution to its mass, not coming from a physical property such as the decay width, but our assessment of the probability of the various theoretical "scenarios". I suggest that this activity is not itself theory, but is one of the things of theory.

EPR and the Dangers of Cannabis

> I have given you an argument;
> I am not obliged to give you an understanding.
>
> *Dr. Samuel Johnson.*

6.1 Cannabis

A common error in the use of statistics is to assume that the existence of a correlation between two random events, revealed by many trials, implies a causal relation. Let me illustrate the truth of this with a story. Suppose "researchers" in a certain state in the US find out, for each year since independence, the number of murders in the state, and also the number of Bibles sold. Suppose that these numbers vary from year to year in an unpredictable way, and so are described by two random variables, X, Y. The sample space is taken to be the integers between 1776 and 2004. From the data we can find the means of X and Y, their covariance, and their correlation coefficient. Suppose we find that the correlation coefficient is 0.8, so that when the murder rate is large, so is the rate of buying Bibles. Can we conclude that (1) a high murder rate in a given year causes people to rush out and buy Bibles? Or, can we suspect that (2) in years when many Bibles were sold, a lot of people read the gory stories in the Old Testament, and decided to follow suit and do some murders of their own? Or, should we (3) reject that the correlation shows any causal relation in either direction? We might look for a common cause for both sets of observations, and might find that both the murder rate and the sales of Bibles are directly proportional to the population, described by a random variable Z, such that $X = aZ + R$, $Y = bZ + S$, where a, b are positive constants (sure functions) and R, S are small random variables. If the latter theory turns out to be true, then someone knowing this explanation of the data would laugh at theories (1) and (2).

A more elaborate example is the classical version of the Einstein-Podolski-Rosen "paradox". Suppose that in a game of cards, the dealer C, sits in London, while Alice is in Paris, and Bob is in Chicago. It is agreed that C will select a pair of cards, one red and the other black, every week. He will toss a coin to decide which card to send by post to Alice; he also sends the other one to Bob. When a card arrives in Paris, but before opening it, Alice knows that there is a fifty:fifty chance that the envelope contains a red card. The same is true of Bob. When Alice opens hers, she immediately knows, not only her card, but also Bob's. Suppose Alice's card is red, and Bob does not open his letter just yet. Bob thinks his chance of a red is still 50 % (and he is right; Alice's viewing of her card does not alter Bob's assessment of his chances); thus there are two different views on the probability that Bob's card is red. We have a game with two players. It would be absurd to claim that the act of seeing her card *caused* Bob's card to flip to black. All Alice did is to reveal events that had already happened. The colour of A's cards are 100 % anti-correlated with B's, because they have a common origin (London) in which they can be chosen to be so: C's coin toss, leading him to choose red for Alice, also *caused* him to choose black for Bob. But Alice is innocent!

A discussion of a similar set-up, but with black and white balls instead of cards, and without the game theory, was presented by Penrose in [133], page 363. He goes on, (page 364) "No-one in his ... right mind ... would attribute the sudden change of the second ball's 'uncertain' state to being 'black with certainty', or to being 'white with certainty', to some mysterious nonlocal 'influence' travelling to it instantaneously from the first ball the moment that ball is examined". This is subtly different from our conclusion, in that Penrose attributes the uncertain state to the second ball and not to the second observer; he is thus able to live with the idea that the state of the ball undergoes a sudden change (which he concludes did not happen in this, the classical, case). In our version, it is Alice's state that undergoes change when she looks at her ball, and this includes new knowledge about Bob's ball. Bob's ball did not change, and neither did Bob's state. At least, Penrose and I agree that there is no causal connection behind the observed correlation.

This subtle inaccuracy in Penrose's wording allows him to continue "Remarkably, ... this viewpoint just *won't work* as an explanation for all the puzzling apparently nonlocal probabilities that arise in quantum theory!" He is one of those people who think that the 100 % correlations predicted by quantum theory between certain sets of observables indicate a nonlocal influence. But we, knowing that the states of the observers have a common origin, and that this is what gives rise to the correlations, just smile.

Here Penrose is not following the Copenhagen view, which states that, when a complete commuting set of observables is measured, then the interpretation of dial readings is as a classical probability model, given by the Gelfand-Kolmogorov construction of sample space and measure. The quantum states (seen by the various observers) are 100 % correlated with the dial readings, so if there is no causal influence in the classical part (the measuring devices) there can be no causal influence in the quantum part: it is not plausible that an atomic spin can cause a distant dial to move.

The question of when and whether a positive correlation shows a causal relation is well illustrated by recent research by the Maudsley Hospital, King's College.

The Government has recently downgraded cannabis from grade B to grade C on the scale of harmful drugs. This is not a very timely decision, in view of some research published in the British Medical Journal and The British Journal of Psychiatry recently. Some earlier (1987) Swedish research [4] using nearly the whole cohort of Swedish males (during national service) showed that there is a correlation between cannabis use in the teenage years and the later incidence of schizophrenia. As explained above, a correlation does not of itself imply a causal relationship. Instead, the Swedish work allows one to speak of an "association" between cannabis use in the teenage years and later schizophrenia. The study showed that the chance of being schizophrenic at a later date was 6 times as high in the group using cannabis than in the group not using it. The pro-cannabis lobby has criticised this and other studies, saying that the analysis does not ask whether there is a common cause of both the psychosis and the use of cannabis; for example, its use was illegal, thus causing a lot of worry. Another suggestion is that in the early stages of the disease, cannabis is perceived to be a palliative, and [seems to] reduce the symptoms, [possibly] leading to more use by those with mental problems than by the general population. The new work [8] tends to discount the theory about illegality, since there was not the same association between schizophrenia and other even more illegal drugs. In the Swedish work, it was not shown that both cannabis use and later psychosis are not both caused by another circumstance, namely, a predisposition to schizophrenia. In [8], this predisposition was measured at age 11 using standardised criteria; this vulnerable group had a statistically significant increased association with later schizophrenia if they took cannabis, as compared with vulnerable children who did not take it; the remaining group, of nonvulnerable children, also showed an increase, but it was less marked. The third result of [8] was that the chance that a cannabis user would develop schizophrenia at 26 was greater for those who started heavy use of cannabis at 15, as compared with a start at 18. The conclusion of the study was that policy should concentrate on trying to delay the start of the cannabis habit in all children, but especially in the vulnerable group.

In [9], a study is made of whether cannabis use is a **cause** of schizophrenia, which could then be regarded as an **outcome**. Here, the three most important items which must be present, if we are to say cannabis is a cause, are association, temporal priority, and direction. I can't do better than quote from [9]: "Association is the requirement that a cause and an outcome appear together. When the putative cause is present, the outcome rate [must be] higher than when the putative cause is absent. There is no requirement for the putative cause to be present in every case of the outcome, just that the rate of outcome is higher in those with it than without it." Thus, association is established if the two effects are positively correlated. More, "temporal priority is the fundamental property that the putative cause be present before the outcome." Finally, "Direction refers to the fact that changes in the putative cause will actually lead to a change in the outcome". The same criterion will be used to show that the correlation in the *EPR* experiment is not causal; there are no influences faster than light in the experiment: Alice's and Bob's spins are correlated (so they have an association) but the requirement of direction does not hold: changing Alice's spin value from the measured value to its opposite does not affect Bob's spin. By the way, the criterion of temporal priority also fails in the *EPR* experiment, since if Bob has already measured his spin, its value cannot be caused by Alice's observation, which comes later.

In [9] the authors do a meta-study of the literature on cannabis and schizophrenia with a view to finding out whether various suggestions are causes of the illness, using the above criteria. The hardest one to prove is direction. In [55] it is remarked that all three criteria have been demonstrated in some cases, and that cannabis intoxication is a **cause** of acute transient psychotic episodes in some people, and that it is a **cause** of recurrence of pre-existing psychotic symptoms. This was not just an association: direction was proved by giving randomly chosen patients the active chemicals in cannabis, and the rest were given a placebo. A few hours later a doctor, without knowledge of whether the patient had been given the drug or a placebo, looked for symptoms of schizophrenia (as assessed by the doctor using standard criteria, combined with regular reports of the patients themselves on a standard form). The results showed a larger incidence of psychosis in those patients who had been given the active chemicals, compared with patients who had been given placebo. In a discussion in The Moral Maze, on BBC Radio Four, Melanie Phillips tried to make this point, but was out-gunned by Stephen Rose. He said that people who [unlike himself] have no expertise in neural science should not speak on the matter, and that "it is very difficult" to establish a causal relation. He did not say so, but he might have been referring to the fact that association (by itself) is not proof of causality. Unfortunately for his argument, this causal relationship was established in [55], since all three requirements were met. This conclusion is not vitiated by the possibility that use of cannabis caused changes in the brain, and that by stopping the use the patient suffered withdrawal symptoms which led to the observed psychosis. The outcome is the same.

According to [9], controversy remains only as to whether cannabis use actually causes permanent schizophrenia in well people, as opposed to temporary symptoms in already ill patients. The authors define "heavy use" of cannabis to be 50 or more joints per year, and they study the already published literature to see whether heavy use by well people is a cause of schizophrenia in the long term. They conclude that association and time priority were established in all the studies they quote. An association persists when other possible causes are factored out, such as use of other drugs, disturbed behaviour, low IQ, place of upbringing, cigarette smoking, poor social integration, sex, age, ethnic group, level of education, unemployment, and single marital status.

The idea that cannabis causes long term psychosis in healthy people is supported but not proved by the relation between the size of dose [of cannabis] and the severity of the disease later in life. The causal hypothesis passes all tests of association and time priority when these other possible factors are adjusted for. This robustness is not true of some of these other possible causes (in particular, use of other drugs): they fail to satisfy association when cannabis is factored out. A history of previous psychotic symptoms is the exception; it is reasonable to assume that such previous history is a cause of some of the later illness. This group [the vulnerable group] is the most at risk of enhanced illness in the group that took cannabis: there, the association is stronger. They conclude that a doubling of overall risk is associated with heavy use, after factoring out the effect of previous psychotic symptoms. The authors predict an increase in schizophrenia in proportion to the increased use of cannabis in the USA and London over the last 10 years. I note also a claimed recent drop in cannabis use in Holland, so another test of association is that the illness should fall there, in line with this reduction, over the next ten or fifteen years. Neither of these predictions, if fulfilled, will be proof of direction: an increase in both cannabis use and in schizophrenia in London and the USA might have a common cause [eg an increase in a population susceptible to both]. Similarly, a decrease in schizophrenia in Holland ten years after the recent observed reduction in cannabis use might, if found, also have a common cause [eg a national decrease in worrying]. However, the recent reduction in the use of cannabis in Holland might have more to do with the police decision to close half the cannabis cafés than a national decrease in worrying. If this can be shown to be the case, and a significant reduction of schizophrenia occur in ten years' time, direction would have been proved. One could then talk about the reduction in illness as having been **caused** by the closure of the cafés. However, no results of this nature have been published as far as I know.

There are several anecdotal reports of immediate and severe schizophrenia occurring after the ingestion of cannabis; thus it is worthwhile studying whether eating it is more dangerous than smoking it, as well as settling the open question of whether cannabis causes permanent schizophrenia in healthy people. Either experiment would be difficult to do; it would need to be a double blind experiment on a random group of healthy people. For example, to test the second hypothesis about healthy people, they should be randomly divided into two groups, and one group are injected a placebo and the other group are injected the equivalent of one spliff, every week for 15 years; then one asks whether the incidence of suicide, religious mania and hallucinating is significantly larger in the drug group than in the placebo group.

The most difficult thing about this is getting the proposal past the King's College Ethics Committee.

The Home Secretary, Mr. Charles Clarke, has since recommended that the downgrade be looked at again, in view of some recent new experiments, and the availability of skunk and other more powerful varieties of cannabis. His conclusion was that he thought that the grading C was to be left in place.

6.2 The *EPR* Experiment

The original paper by Einstein, Rosen and Podolski [62] argued that quantum mechanics was incomplete, since, it was claimed, quantum theory did not assign a real value to each 'element of reality'. *EPR* state that "if, without in any way disturbing the system, something can be measured, then it is an element of reality". They illustrated this with a spin-zero atomic system at rest, decaying into two particles of the same mass. By the law of conservation of momentum, they must fly off in opposite directions, each direction being equally likely. When they are far apart, we measure the momentum of one of the particles without disturbing the other, by the principle of causality in special relativity. The momentum of the other particle is now known to us; it is the negative of the observed value for the first particle. We can now measure the position of the second particle, [assuming that this causes no change in its momentum] and thus know both its position and it momentum. Thus the point in phase space occupied by the particle is an element of reality. *EPR* define a theory to be *complete* if it assigns a real value to each element of reality. Since in quantum mechanics it is impossible to assign both the position and momentum to a particle, they conclude that quantum mechanics is not complete.

The original set-up was modified by Bohm, in order to avoid unproductive arguments about the rigour of sharp eigenstates of operators with continuous spectrum. In Bohm's version, the spin zero atomic system at rest decays into two particles of spin $1/2$, which move off in opposite directions. Since angular momentum is also conserved, the spins of the two particles are opposite. If we now measure the spin S_z of one of the particles in the z-direction, say we get $+1/2$, then we know the spin of the second particle in the z-direction; it must take the opposite value, $-1/2$. Since we can wait until the particles are space-like separated before doing this, we do not disturb the second particle. An observer at the second particle can now measure the spin of the second particle in the x-direction, and will get a value. Thus we know the spin of the second particle in both the z-direction and the x-direction, so both are elements of reality. But quantum mechanics does not assign them simultaneous values, and so is incomplete. I now argue that, contrary to the belief of its authors and many since, this ruse does not achieve the stated aim.

The definition of "element of reality" given here leaves much to be desired. That we can find the momentum, or spin, of one of the particles by measuring that of the other is possible only because of the special nature of the entangled state. In modern terms, we do not require that some observables can be

measured in certain states and not in others. We can and do measure the
momentum and spin even if it does disturb the system. However, *EPR* do
claim to have constructed special circumstances in which the measurement of
non-commuting observables has been achieved, whereas quantum mechanics
cannot describe any such circumstance, and so is incomplete. Unfortunately,
the argument of *EPR* is fallacious: neither of the two space-like experimenters,
Alice and Bob, can know the result of the measurement by the other. Quantum theory with two observers involves two distinct algebras of observables,
one for each observer. In this case, these algebras commute, and measurement
goes ahead by A and B without either disturbing the other. Neither can find
a value for both S_z and S_x for the particle in hand. Nor can a third collab-
orator, looking at the quantum record later, assert that either particle has
a definite value for both S_z and S_x. Suppose that A finds that $S_z = +1/2$
for her particle and that B finds that $S_x = +1/2$ for his particle. The same
remains true after both measurements, as neither affects the other. This is
consistent with the von Neumann measurement: let P be the projection onto
the eigenstate $S_z = 1/2$, and Q the projection onto the eigenstate $S_x = 1/2$.
Then the final state, including the conditioning due to observation, is

$$\Delta_{\rho_f} = \frac{(P \otimes Q)\Delta_\rho(P \otimes Q)}{\text{Tr}[(P \otimes Q)\Delta_\rho(P \otimes Q)]}. \tag{6.1}$$

Here, Δ_ρ is the density operator for the entangled state:

$$\Delta_\rho = 1/2\left\{ \begin{pmatrix} 1 & 0 \\ 0 & 0 \end{pmatrix} \otimes \begin{pmatrix} 0 & 0 \\ 0 & 1 \end{pmatrix} - \begin{pmatrix} 0 & 1 \\ 0 & 0 \end{pmatrix} \otimes \begin{pmatrix} 0 & 0 \\ 1 & 0 \end{pmatrix} \right. \tag{6.2}$$

$$\left. - \begin{pmatrix} 0 & 0 \\ 1 & 0 \end{pmatrix} \otimes \begin{pmatrix} 0 & 1 \\ 0 & 0 \end{pmatrix} + \begin{pmatrix} 0 & 0 \\ 0 & 1 \end{pmatrix} \otimes \begin{pmatrix} 1 & 0 \\ 0 & 0 \end{pmatrix} \right\}. \tag{6.3}$$

A direct calculation shows that the final state, Δ_{ρ_f} of eq. (6.1) is $P \otimes Q$.
If it turns out that Alice did her measurement first, then before Bob did his
measurement, Alice's state has density matrix

$$(P \otimes I)\Delta_\rho(P \otimes I)/\text{Tr}[(P \otimes I)\Delta_\rho(P \otimes I)] = P \otimes (I - P).$$

Thus Alice is supplying Bob with the eigenstate of S_z with eigenvalue $-1/2$.
This is so whether Alice tells Bob her result at the time, or later shows him her
lab book. Bob subsequently makes a measurement of S_x. This measurement
disturbs the eigenstate $S_z = -1/2$, and nobody has a particle at hand with
sharp values of both S_z and S_x. Similarly, if Bob did his experiment first, he
supplies Alice with the eigenstate of S_x with eigenvalue $-1/2$; this is disturbed
when she measures S_z.

EPR say "... one would not arrive at our conclusion [the incompleteness]
if one insisted that two or more physical quantities can be regarded as simultaneous elements of reality *only when they can be simultaneously measured or
predicted*". They were, then, aware that Bob must wait some time to receive

the result of Alice's measurement of S_z. Bob would not be wrong in assuming that the value of S_z for his particle would remain unchanged, being opposite to Alice's, during his wait, and that he would know it when her message arrived. *EPR* thought that he could then measure S_x and end up knowing both; quantum theory, and experiment, says not.

Thus, it is my opinion that the authors of *EPR* wrongly thought that there was an *experimental* situation in which a particle could be found to have sharp values of position and momentum. They might have also had in mind a *doctrinal* objection to uncertainty, but I do not think that this is relevant. The error (in Bohm's version) is to assume that S_x of Bob's particle is a real number, rather than an operator, before Bob measured it; or that after Bob's measurement of S_x, that his particle's S_z still had the same value it had before the measurement. That is, they were arguing as if there were a sample space for the system.

In a relativistic theory, we adopt Haag's algebraic theory, and postulate that observables are localised in some bounded region of space-time. That is, to each bounded set $\mathcal{O}$ of space-time is given a C^*-algebra, $\mathcal{A}(\mathcal{O})$ whose hermitian elements represent observables that can be measured by an observer, Alice, in $\mathcal{O}$. Let $\mathcal{A}$ be the C^*−algebra generated by all the local algebras for all bounded open sets $\mathcal{O}$. Let ρ be the state, and $\rho|_{\mathcal{O}}$ denote the restriction of ρ to $\mathcal{A}(\mathcal{O})$. This will be represented by the density operator Δ_1 acting on the Hilbert space of states in the sector. Any $A \in \mathcal{A}(\mathcal{O})$, has spectral resolution $A = \int a \, dP(a)$. Consider the spectral projection

$$P_i := \int_{a_i}^{a_{i+1}} dP(a)$$

onto the spectrum of A in an interval $[a_i, a_{i+1}]$. These lie in $\mathcal{A}(\mathcal{O})$ and the Lüders map

$$\Delta_1 \mapsto \sum_i P_i \Delta_1 P_i$$

represents the change in the local state due to the measurement of which interval the observable A lies in, when the state is ρ. This map is completely positive and stochastic, and maps the local algebra to itself. Consider now a relativistic theory. Each Lorentz transformation L takes a point $x \in \mathbf{R}^4$ to the point $Lx \in \mathbf{R}^4$; applied to each point in the region $\mathcal{O}$ we get the transformed region $\{y : y = Lx, x \in \mathcal{O}\}$; this will be denoted $L\mathcal{O}$. The assumption of Lorentz symmetry then ensures that there will be a unitary operator $U(L)$ which maps the algebra $\mathcal{A}(\mathcal{O})$ by conjugation to $\mathcal{A}(L\mathcal{O})$. This is seen to map the Lüders map for measuring A into the Lüders map for measuring $U(L)^{-1}AU(L)$, and transforms the spectral projections and so also the spectrum of A in some way. This is the way the dial showing the result of the measurement must transform; this is how the dial of an observer related to Alice by L should be calibrated.

If now we have another observer, Bob, in region $\mathcal{O}'$ space-like to Alice, he will have his own algebra $\mathcal{A}(\mathcal{O}')$, which is postulated (by Haag) to commute with $\mathcal{A}(\mathcal{O})$. Bob will see the state with density operator $\varDelta_2$, corresponding to the state $\rho|_{\mathcal{O}'}$. Suppose he measures the spectral projections of B with the Lüders operator $\varDelta_2 \mapsto \sum_j Q_j \varDelta_2 Q_j$. Now the argument goes as in the non-relativistic case, given above. Alice's measurement is the identity map on Bob's algebra, and *vice versa*. There is no transmission of information faster than light, and we see a direct relationship of this fact to the commutativity of observables localised at space-like separation. This property is called "Einstein causality" by Haag, in recognition, perhaps, to Einstein's contribution to *EPR* and his earlier remarks on quantum measurement.

Subsequent writers changed the paradoxical element in the *EPR* experiment, with less emphasis on incompleteness. Instead, it is supposed to show that quantum measurement has nonlocal aspects. Thus, Michael Brooks writes [35] "... two particles become 'entangled'. If you then change the spin of one particle it will instantly affect the spin of the other, regardless of the distance between them". Here, the most direct meaning of the phrase, 'change the spin of one particle', is that the spin had one value, and was to be changed to another. If this did, indeed, instantly affect the spin of the other, then to the property of correlation between the spins [due to the entanglement] can be added the properties of *direction* and *time priority*. Then we have all that is needed to be able to assert that the value of the spin of the first particle is one of the causes of the value taken by the other. There would be faster-than-light transmission of information. Of course, this does not happen; after A has measured S_z and obtained $+1/2$, any later rotation of A's spin to another value does not affect the fact that if B measured S_z he would find $-1/2$. Brooks's article needs a more careful interpretation, or better, a complete revision; it is a lost cause.

A more cagey statement of the claimed nonlocal effect of a measurement is to be found [133], p. 370, where it is written "... the nonlocal 'influences' that arise in *EPR*-type experiments are not such that they can be used to send messages .... It is of no use to be told that a photon is polarised 'either vertically or horizontally' (as opposed, say, to 'either at 60° or 150°') until one is informed *which* of the two alternatives it actually is. It is the first piece of 'information' (i.e the *directions* of alternative polarisation) which arrives faster than light ('instantaneously') while the knowledge as to *which* of these two directions it must actually be polarised in arrives more slowly ..." [Note that for photons, vertical or horizontal refers to the direction of the electric field, not the direction of spin].

Penrose is almost saved from error by the use of inverted commas around the word 'information'. Indeed, if the spin-state of the entangled state is given by eq. (6.3), then the state $\rho|_{\mathcal{O}'}$ as seen by Bob is the unpolarised one. This can be written equally as the half-and-half mixture of spin up and spin down, or as the half-and-half mixture of spins in any other two opposite directions. Thus Bob already knew that his state was the equal mixture of any two opposite spin directions and there was no information in the first piece of 'information'. Thus the quoted passage, while true, has no content; it does not tell us what these supposed nonlocal influences are. If on the other hand, Penrose intends the reader to think that Bob gets instantaneous knowledge of what spin-direction Alice is measuring, then the passage is in error. First, because this *is* real information and can be used to send messages, and secondly, because Bob cannot get such information instantaneously. It is curious that Penrose is nearly saved by the fact that the state space in quantum mechanics is not a simplex, and that the decomposition of a mixed state into pure states is not unique. The density matrix is all there is: two mixtures with the same density matrix cannot be distinguished. Some authors have espoused the idea that there *must* be some remaining information which can be used to distinguish two mixtures of different pure states to form the same density operator. Then the state space is not the dual of the observable algebra, but the set of probability measures on the boundary of the dual, the pure states. This is the formal way to put the idea that the set of pure states or normalised wave-functions modulo phase behaves as the quantum analogue of phase space, the set of $(\mathbf{x}, \mathbf{p})$, which plays a role in classical Hamiltonian dynamics. This does not lead to quantum theory, however. For there are states which are the same in quantum mechanics which differ in this version. To see this, let ψ_+ and ψ_- be normalised eigenstates of σ_z with eigenvalues respectively 1 and -1, and let φ_+ and φ_- be normalised eigenstates of σ_x with these eigenvalues. Each of these is a two-component column matrix. Then the expressions

1. $\psi_1 := \psi_+ \otimes \psi_- - \psi_- \otimes \psi_+$
2. $\varphi_1 := \varphi_+ \otimes \varphi_- - \varphi_- \otimes \varphi_+$

are both proportional to the singlet state of a system of two spin $1/2$ particles, and so are the same in quantum mechanics. This is seen easily by expanding $\varphi_\pm$ in (2) in terms of $\psi_\pm$. They would be different in the version in which one could distinguish the way the state has been made. If this part of this information could be sent instantaneously, so would Penrose's 'information' that he, Bob, had received the state ψ_1 rather than φ_1 say, from Alice, and messages could travel faster than light, contrary to experiment. For this reason, 'projective Hilbert space as phase-space' is a lost cause.

On page 371, Penrose discusses two observers C and D, moving relative to each other, and claims that they have two inconsistent pictures of the measurement of an entangled state. Let Δ be the density matrix of this state. Penrose does not have observers Alice and Bob, measuring the spins of their respective particles, but instead talks about the left-hand part of the state and the right-hand part of the state. Suppose that both A and B measure the z-component of spin. Suppose that both the observers, C and D observe the state of the whole system of two particles over time, as if they had access to the lab-books of both Alice and Bob. Penrose claims that the measurement of the spin on the left "causes the nonlocal 'jump' " of the right-hand part of the state. Then, in the set-up given, this collapse of the wave-packet on the right appears to D to have happened *before* Bob measured it.

This is an easy one; I see no evidence for any cause. Association is present, but direction does not hold, as we saw above in the paragraph about Brooks. Time priority also does not hold in all frames. One observer, C, notes that in his Lorentz frame the time at which Alice measured her spin was before the time that Bob measured his. C might notice the resulting collapse allowed Alice to predict what Bob's measurement would reveal. The other observer, D, sees that, in his frame, Bob made his measurement before Alice, and so D might remark that Bob could predict her result. If the frames of Alice and Bob are different from those of C and D, then they might order the measurements differently from C or D. Instead of predicting what Bob's result will be, Alice might find that she can say what it was. I see no inconsistency here: neither measurement causes the other, but merely reveals the result. Since all observations commute, the physics is isomorphic to the case of sending letters to Paris and Chicago. In his version of the letters, p. 363, Penrose does not suggest that the classical problem has a similar inconsistency when we make it conform to special relativity. The concept of consistent histories, which will be described in the next section, puts it like this: each version, seen by C and D, is a consistent history of the events; we must distinguish between true statements and statements that are only "reliable". C can say that Alice made a measurement showing that her spin S_z was $1/2$, this is true. C can then conclude that just before Bob made his measurement, his spin S_z was $-1/2$. This is a consistent, or reliable, statement, but is not true; there is no pointer in Bob's record showing this before he did his experiment. Similarly, in D's frame, the result of Bob's measurement, $S_z = -1/2$, is true, but his prediction of Alice's result before it has been performed is merely consistent.

On p. 386 of the book [133], Penrose claims that quantum mechanics gives the wrong answer to the cat paradox, and other classical objects; he puts forward the idea that quantum mechanics requires there to be a complex phase between different classical states, contrary to experience. The word *superselection rule* does not appear in the index, but he seems to be feeling towards the possibility that the correct theory should give classical mechanics for large objects and quantum mechanics for small objects. If so, then welcome, Roger, to modern quantum mechanics, which is von Neumann with superselection rules and local observers. There is no need for a new theory. Classical mechanics emerges as the large-scale limit of quantum theory. Our view of the paradoxes leaves Penrose with one problem: where do we switch from quantum to classical? The answer is, for systems without macroscopic quantum effects, we can put the cut anywhere, for all practical purposes. The justification of this statement can be found in the deep results on the thermodynamic limit of quantum systems, where the *observables at infinity* are proved to be classical variables, which are superselected. For any large, finite version of such a model, the quantum phases between the putative classical variables average out and the thermodynamic phases, while not mathematically disjoint, last for millions of years. A detailed study of why classical objects have no phase relations between them has been given by Omnès; a popular word for it is decoherence.

There are also dynamical models [87, 88] of the measurement process, which show nicely that as the measuring device gets larger, but is still finite, then the complex phases between the pointer states are non-zero but are getting smaller, so that in the limit the pointer readings become classical. It is this limit that we call an ideal measuring device. Hepp's work has been criticised by Bell [20], who argues that Hepp needs to choose his observable first, and take the limit later. Bell argues that for a measuring device of any finite size, one can find an observable without a classical value for the chosen measuring device. While true, this argument does not invalidate Hepp's work; the needed observable might well not be easy to find. Omnès gives a model calculation, mentioned in Chap. (4), that to measure such an observable would need the system plus observer to be far larger than the present universe.

6.3 Counterfactuals

The *EPR* set-up has been developed into more complicated situations such as that of Greenberger, Horne and Zeilinger [76]. This work will be denoted by *GHZ*. These are new hoops for us to jump through; when we have done this, Penrose [134] has concocted indefinitely complicated examples of similar "paradoxes". But hard cases make bad law, and I shall limit myself to the original idea of *GHZ*. There, there are three observers, mutually space-like separated, each receiving one of three particles, each with spin 1/2. These particles arrive at the observers after the decay of a three-particle unstable state in the pure state $\psi := 2^{-1/2}(|+++\rangle - |---\rangle)$, where $\pm$ denotes that the spin is $\pm 1/2$. Let

$$a = \sigma_1, \qquad b = \sigma_2, \qquad c = \sigma_3$$

denote the three Pauli matrices, and let a_i denote the measurement of σ_1 of the ith particle by the ith observer, $i = 1, 2, 3$. Similarly, b_j and c_k denote the measurements of σ_2 and σ_3 by the jth and the kth observer, and also the values obtained in the measurement. Then in the state ψ, one can show that with probability 1,

$$a_1 b_2 b_3 = 1; \qquad b_1 a_2 b_3 = 1; \qquad b_1 b_2 a_3 = 1.$$

In each case the three operators in the relation commute, and although these are not operator relations, and the values observed in the state ψ can be either $+1$ or -1 for each observable, they are strongly correlated and the products will always be observed to be 1. Now suppose that b_1, b_2, b_3 are measured. We see that if b_1 and b_2 are known, (by observers 1 and 2 sharing their information later) then 1 and 2 could predict what a_3 would have been if 3 had measured a instead of b. This is a counterfactual; since 3 measured b, not a, it did not happen. But it is a mild sort of counterfactual: 3 could have measured a, and quantum mechanics predicts that, if 3 had measured a, then the prediction made by 1 and 2 would be correct. Similarly, if 2 and 3 get together, they can predict what 1 would have seen if she had measured a instead of b, and moreover, this prediction is 100% verified whenever it is tried out. Similarly, if 1 and 3 get together, their measurements of b_1 and b_3 correctly predict what 2 would have got for a_2 if he had measured it instead of b_2. Now, the result got by 1 and 2 would have been the same if 3 had measured a instead of b, since they are far away. Similarly, for the other pairs of observers. Suppose now that b_1, b_2, b_3 are all equal to 1 (which happens sometimes). Then the three observers can later get together in the three pairs (12), (23) and (31) and say that if they had all measured a instead of b, they would all have found the result 1. Then they would predict that $a_1 a_2 a_3 = 1$. But this is not true: this value for the product has probability zero in the state ψ, and the value -1 has probability 1!!

This paradox is understood by the remark that there is no overall sample space in quantum mechanics; each complete commuting set, and state, defines a sample space and a probability on it. The classical description in terms of values and probabilities is *contextual*, not absolute. We cannot say that there is a sample, and all random variables take a value at the sample. When 3 considers measuring a instead of b, the possible results for a are not events in the sample space already created by the measurement of b. They are events in another sample space, which was not created by the actual measurements made. Such an event is called a counterfactual. We have seen that it is sometimes consistent to allow one counterfactual, that 3 measures a instead of b; but we get a paradox if we allow three counterfactuals. In fact, the argument breaks down when two counterfactuals are considered. If 2 and 3 had both measured a, then no conclusion with probability 1 can be drawn by any of the three pairs of observers.

There is no paradox if we introduce information algebras for the individuals: each measures b and does not even know what the others are measuring, let alone what the result was or will be. The observers can share information; if 1 and 2 share that they both measured b and got the result 1, then they can correctly predict what 3 would get if he measured a, namely, 1. When they learn that he measured b, this prediction has no permanence if further counterfactuals are added. This point was made by Griffiths in the "consistent histories approach" to quantum mechanics, which is a version of the Copenhagen interpretation in which all information is recovered by all observers. We now describe this idea.

When we consider observing a single observable X, we can always describe the results by using a classical probability space, whatever the state ρ happens to be. Of course, this just uses Gelfand's probability space, since the algebra of polynomials in X is abelian; the measure to use is then determined by ρ. See Sect. (3.6). From the present point of view, the probability that $X = \sum_{j=1}^{n} x_j P_j$ take the value x_i is

$$p_i = \mathrm{Tr}\,(P_i \rho P_i) = \mathrm{Tr}\,(\rho P_i),$$

and this is linear in P_i. Now, the projection onto the set spanned by the eigenvalues $x_1, \ldots, x_s$ is $P_s = \sum_{j=1}^{s} P_j$, and the probability of getting this is

$$p_s = \sum_{j=1}^{s} \mathrm{Tr}\,(\rho P_j) = \mathrm{Tr}\,(P_s \rho P_s), \tag{6.4}$$

where we have used that $\mathrm{Tr}\,(P_i \rho P_j) = 0$ if $i \neq j$. This is what leads to the classical description.

However, if we contemplate the sequential measurement of X and Y which do not commute, then we cannot in general write the probability that we find a given collection of the eigenvalues of X and Y in a similar form. The failure of X and Y to commute might arise if X and Y represent that same observable at different times; Griffiths takes this to be the case. The general case is entirely similar to his results. He asked, when we allow all the options for the results (x_i, y_j) in the two measurements, when can we say that the history has a classical description? For example, take $X_1 = X(t_1)$, $Y = X_2 = X(t_2)$, with $t_1 < t_2$. Write the spectral resolutions

$$X_1 = \sum_{i}^{n} x_i P_i(1), \qquad\qquad X_2 = \sum_{i}^{n} x_i P_i(2).$$

We may take it that $[P_i(1), P_j(2)] \neq 0$. A *history* is then a pair (x_i, x_j). At each time, we have completeness and mutual exclusivity:

$$\sum_{i} P_i(k) = 1, \qquad k = 1, 2$$

$$P_i(k)P_j(k) = \delta_{ij} P_i(k), \qquad k = 1, 2.$$

Then we know that the probability of the history (x_i, x_j) arising is

$$p_{ij} = \mathrm{Tr}\,(P_j(2)P_i(1)\rho P_i(1)P_j(2)). \tag{6.5}$$

Although we have $\sum_{ij} p_{ij} = 1$, the p_{ij} are not joint probabilities of two classical variables; for example, x_i arose earlier than x_j, and $p_{ij} \neq p_{ji}$ in general. Griffiths asks, on what conditions on ρ allow us to make a consistent classical theory out of the set of histories? Certainly, this should be possible if X_1 commutes with X_2. For then we can simultaneously diagonalise $P_i(t_1)$ and $P_j(t_2)$ for all i and j, leading to $p_{ij} = p_{ji}$ being true for all i and j.

The probability that we have the two histories (x_i, x_j) and $(x_{i'}, x_{j'})$ is just $p_{ij} + p_{i'j'}$ and this gives

$$\begin{aligned}
p_{ij} + p_{i'j'} &= \mathrm{Tr}\,[P_j(2)P_i(1)\rho P_i(1)P_j(2)] + \mathrm{Tr}\,[P_{j'}(2)P_{i'}(1)\rho P_{i'}(1)P_{j'}(2)] \\
&= \mathrm{Tr}\,[(P_j(2)P_i(1) + P_{j'}(2)P_{i'}(1))\,\rho\,(P_i(1)P_j(2) + P_{i'}(1)P_{j'}(2))] \\
&\quad - \{\mathrm{Tr}\,[P_{j'}(2)P_{i'}(1)\rho P_i(1)P_j(2)] + \mathrm{Tr}\,[P_j(2)P_i(1)\rho P_{i'}(1)P_{j'}(2)]\}\,.
\end{aligned}$$

If X_1 commutes with X_2, then the last two terms vanish, because of either $P_i(1)P_{i'}(1) = 0$ or, if $i = i'$, we have $j \neq j'$ and so $P_j(2)P_{j'}(2) = 0$. For certain ρ, the last two terms cancel even in the non-commutative case; Griffiths saw that this allows one to speak of a classical history. The condition for the cancellation is that

$$\mathrm{Re\,Tr}\,[P_{j'}(2)P_{i'}(1)\rho P_i(1)P_j(2)] = 0 \qquad (6.6)$$

for all $(i,j) \neq (i',j')$. This is the *consistency* condition of Griffiths. It has been used by Gell-Mann and Hartle in quantum cosmology, where it is approximately proved by the phenomenon of decoherence. A similar condition was given by Griffiths [78, 79] so that a finite sequence of non-commuting operators $X_1, \ldots, X_m$ can be treated as a classical process. The book of Omnès [131] has a clear description of this. Thus, the problem as to why large quantum systems behave classically is understood.

The continuing work on quantum paradoxes parallels a similar output after relativity was first formulated: when simple paradoxes concerning time-dilation or Lorentz contraction were explained, more and more elaborate versions, (twin paradox, thermodynamic paradoxes involving absorption and emission in a gravitational field ...) were claimed to show some mathematical or physical inconsistency in the special or general theory. Rebutting these took up the time of the advocates of the theory, and led to clarification. But the activity died out. The quantum paradox community is, even after all these years and all the rebuttals, still on the increase. In has spawned larger enterprises, such as Bohmian mechanics, Nelson dynamics, geometro-stochastic dynamics, quantum-state reduction[135], and the quantum theories of the brain of Penrose and of Stapp. These are lost causes too, and fall down with the rebuttal of the spooky action at a distance claimed to inspire them.

7

Stapp's Theory of the Brain

> "In the great drama of existence we ourselves are both
> actors and spectators"
>
> Niels Bohr [30, 31]

This chapter is a version of a critique of a website entitled "Why classical mechanics cannot naturally accommodate consciousness but quantum mechanics can", by Henry P. Stapp [153]. I called my article "Quantum theory on the brain", [160]. Stapp's webpage was a summary of his book [152].

We argue that Stapp has made three errors, one in each of the subjects of classical mechanics, quantum mechanics and experimental brain dynamics. Together they make his version of brain theory a lost cause.

7.1 The Speed at Which Concepts are Formed

Stapp wants a model of the brain which 'can bring the information together from far-apart locations' instantaneously to get the 'whole thought'. However, whole thoughts are **NOT** arrived at instantaneously; this is well known; see for example, [49]. By measuring brain activity and asking the subject what he is thinking, it can be seen that one collects one thoughts from various parts of the brain, and that this activity starts about 1/2 second before the conscious part of the brain is aware of the thought or decision. We do not need any information to flow at the speed of light to explain these facts; we certainly do not need any effects that travel faster than light. Yet this is put as one of the failings of classical mechanics: it does not allow faster than light flows of information. Thus, usual theories of the function of the brain are quite adequate, at least on this question.

7.2 The Claimed Absence of Correlations in Classical Field Theory

He first (in 2.6 of [153]) tries to model the brain as the aggregate of a large number of spatially localised computers, each in some configuration. It is then argued that it is not possible to follow the dynamics of a brain by using such a model. The (discrete-time) dynamics is given by local update-rules; the new value of the configuration at a point involves the state at the point and the neighbouring points. There is no interaction taking place instantaneously over long distances. Thus his idea here is to describe the brain as a random field over a lattice; the values taken by the field are discrete. He says in (2.7), '...the information stored in any *one* [original italics] of the logically independent computers ... is ... minimal: it is no more than is needed to compute the local evolution. This is the analogue of the condition that holds in classical physics'. He seems to be saying that the information held at a point x by the model is just enough to calculate the probabilities of local changes needed to specify the time evolution. This is a deficient description of classical physics. Firstly, he does not distinguish between the fields (the observables) and the states (the probabilities that a certain configuration is realised). Indeed, it is clear that he does not allow a probabilistic description of the state, even though the brain is very noisy, and all the leading models in the subject of neural nets use (classical) statistical physics. But if his description is non-random, what does he mean by saying that the computers are 'logically independent' of each other? Recall that there are three distinct meanings that the word "independent" might have, as I have explained in Chap. (4). Does he mean dynamically independent, or even 'statistically independent'? He gives a hint. In (2.4) he uses the analogy of a display of pixels on a TV screen as the state of the computer. He writes that his requirement, termed 'minimal information', gives rise to an 'intrinsic description without any explicit relationship that might exist between the elements, [such as] that Pixel 1000 has the same colour as Pixel 1256'. Now, in a non-random description by a configuration, the state of each pixel is determined, as is the question whether it has or has not the same value as another specified pixel. If this information is missing, it must be because the writer has lost it, perhaps by replacing the (pure) state by a mixed state, the product of probability measures at each site. For such a state, one may indeed say that the computers are statistically independent, and that the state shows no correlation between different sites. It is the presence of such correlations in quantum theory that entices the author to invoke quantum theory. However, the author would be quite wrong to suppose that classical field theory cannot have states with correlations. The original description by pure states has got such information contained in it. But even in a random theory, we can have correlations between the sites, and this will occur if causal effects are present: if a cascade of firing neurons causes others in its path to fire, then at a later time, there will be positive correlations between the firing rates at separated points in the brain. These can be read (slowly) by other parts of the brain, and the information they hold summarised in a new register, forming the 'whole thought'. Correlations have been studied both theoretically and experimentally [138].

The concept that the author describes, a pure configuration of the brain containing no information about the relations between different points, does not exist in classical mechanics.

The author introduces a concept, that of logical independence, which usually has the following meaning: two propositions P and Q are logically independent if it is possible for all four truth values to hold: P is true and Q is true, P is false and Q is true, P is true and Q is false, P is false and Q is false. In the context of lattice field theory, logical independence means that the sample space of the lattice model is the product of the sample spaces at each site, as in Eq. (3.8). It says that any configuration is possible, including one in which two sites have the same value of the field. Thus Stapp's remark about pixels not having the same colour is misleading at best. Statistical independence of the sites is a different concept from logical independence; the sites are statistically independent if and only if the measure is a product measure, as in Eq. (3.9). In most models of the dynamics of classical lattice gases, the measures are not independent at each site, and correlations over large distances are possible.

7.3 Supposed nonlocalities of Quantum Theory

I have explained in Chap. (6), that the alleged nonlocality shown by the *EPR* experiment should be called 'contextuality' in classical mechanics, and that quantum theory does not suffer from this. I argue that when a commuting set of observables is measured, the interpretation of any correlations must be the same as in classical probability theory. There, it is well known that a correlation between two observables does not imply a causal relation in either direction. Stapp contradicts this in his Appendix D4a as follows:

"Orthodox quantum field theory in its covariant form ... is nonlocal in the sense that for certain systems ... the set of correlations predicted by quantum theory ... is incompatible with the following 'locality' condition: the result of any possible measurement *M* must be independent of any free choice that is to be made *later* ... One *cannot* assume that there is no faster-than-light influence of any kind."

This is what is behind Stapp's remark, on p 82:

"... the nonlocal transfer of information apparently demanded by Bell's theorem".

Stapp bases his pessimism on a version of the *EPR* paradox devised by L. Hardy [84]. This makes use of counterfactuals, and the implicit continued existence of a sample point both for the factual and the counterfactual situations. Moreover, Stapp refers to the Bohm model as a nonlocal theory with hidden variables which gives the same results as quantum theory, [152], p. 18. Although this agreement with quantum mechanics was claimed by Bohm and has been repeated by many authors [20, 170, 122], only the statistics of the position variable (at each time) correctly reproduce quantum theory. See Chap. (9). To get the quantum results for the measurement of the position variable at different times, for example, requires that the wave-function be conditioned as in quantum theory after the first measurement. This differs from the classical notion that the measurement only reveals where the particle is.

Again, in Appendix 4b, Stapp writes "It [quantum mechanics] is nonlocal in ... that any ... collapse of the wave-function ... changes expectation values all over the universe". As I have explained in Chap. (4) the act by A = Alice of measuring a local observable (but not yet reading the pointers) has no effect whatever on Bob's state, if his observables are space-like to the measuring apparatus. On reading the pointers, Alice conditions her state by using Bayes's formula. This gives her updated information on her *information algebra*, which is the quantum version of information set used in game theory. Since this information cannot be passed to B until later, it does not affect his state at all. So Stapp has failed to distinguish information available to Alice but not to Bob. The collapse of Alice's wave function is nothing other than the application of Bayes's theorem. To suggest that Bob's statistical description must change just because Alice has some information, is incorrect, and would give the wrong results also in classical game theory. For example, consider a game of cards between two players, in which each player had a hand unseen by the opponent; it is absurd to say that when one of them, Alice, finds that her hand includes the ace of spaces, so the other, Bob, immediately alters his assessment of the likelihood that this is so. Alice's measurement merely reveals the situation to her; it does not cause any change in Bob's hand or strategy. I maintain that information obtained by measurement in quantum theory cannot be passed faster than light, and that information held in various parts of the brain needs time to form a whole thought, decision or idea, just exactly as is observed (especially at my age).

7.4 Conclusion

The three ideas that gave Stapp the motivation to introduce quantum probability into brain dynamics were, I have argued, erroneous. I have not discussed the startling conclusion to Stapp's theory: the brain is able to make a conscious decision to choose one of the outcomes of a quantum uncertainty. He also identifies the creation of ideas with the construction of the events by a measurement and the knowledge gained by looking at the dials (all instruments and dials are in the brain itself). While a description by classical probability allows for the acquisition of knowledge from the state of the brain, it is true that there, in the classical case, one is merely noting an event (such as a shower of firings by neurons) that was there already. We know that without measurements there are no events in quantum mechanics. According to Copenhagen, events are created by measurement. Thus I could take the line that Stapp's fervent advocacy of the Copenhagen view of quantum mechanics (in the original form, without the modifications needed for two observers) is a strong point of the book. One *could* argue that there is a difference between interfering with a classical chance event, which already exists, and merely helping along a quantum event which does not yet exist. To change an event already in the brain needs a physical act, and the decision to implement this is prior to the change, and so cannot be caused by the change. The root of decision-making by the brain might lie in its ability to influence the outcome of a purely classical chance event before "Nature" has chosen the outcome. Now, this does sound spooky; Stapp, who already believes that influences faster than light that cost no energy occur in quantum theory, is prepared to say that something of this sort occurs in quantum systems, but never in classical systems. It is then a matter of almost religious faith to follow Stapp and argue that the mind can choose one or another of the quantum outcomes during the measurement process, and that this free will is not allowed by classical probability.

However, I must remain sceptical that there are any entangled states in the noisy environment of the brain; without these, classical probability will suffice.

8

Hidden Variables

> "[T]here's much more difference ... between a human being who knows quantum mechanics and one that does not, than one that doesn't and the other great apes".

> M. Gell-Mann, Amer. Assoc. for the
> Advancement of Science, 11 Feb. 1992.

This subject has been thoroughly worked out and is now understood. A thesis on this topic, even a correct one, will not get you a job.

In the early nineteenth century, Kant wrote a treatise [98] which argued that Euclidean geometry can be seen to be true by pure logical thought, and does not need any experimental verification. After the discovery of relativity, philosophers have criticised Kant, not so much for not expecting the discovery by mathematicians of non-Euclidean geometries, or for not anticipating the move in philosophy to logical positivism in the twentieth century, but for taking the risky course by debating physics. Physics is an experimental subject, these critics said, and Kant should have confined his comments to a part of mathematical knowledge that is true and fixed, such as probability theory. How wrong were they to choose this topic!

Just as non-Euclidean geometry is a generalisation of Euclidean geometry, so quantum theory is a generalisation of probability theory, and quantum mechanics is a generalisation of a stochastic process. The theory of a stochastic process $X(t)$ was first put on a sound mathematical footing by Kolmogorov in his book of 1933 [108]. By a Kolmogorovian theory, we mean a set-up that conforms to the axioms of this book. Later in life, Kolmogorov played with the ideas of von Mises, who tried to found probability theory on a frequentist philosophy. This programme has since been made completely rigorous by van Lambalgen, who, according to Gill, had to replace the axiom of choice by a new notion. We do not refer to Kolmogorov's late-life conversion as Kolmogorovian in this article. Indeed, we call the frequentist approach "prekolmogorovian". Perhaps a better word is "contextual", as was explained in Chap. (3).

The key to Kolmogorov's theory of a stochastic process $X(t)$ is the proof that there exists a single sample space, on which all the $X(t)$ in the process are random variables. This sample space is the space of paths: at each time, $X(t)$ has a definite value (for each sample), so $X(t)$ is a random version of an "element of reality" in the words of Einstein. He says that, in any theory, a concept (such as that given by a symbol) arising in the theory must take a real value if it is an element of reality. Surely, he meant to allow that it could take different values in different states. We may then take the set of states envisaged by EPR as the sample space Ω, and an element of reality is then a real-valued function on Ω. If we add the technical requirement that the function has to be measurable in the mathematical sense, then the set of elements of reality is a subset of the set of random variables on Ω, and we can use Kolmogorov's theory. Bell's inequalities, which hold in any Kolmogorovian theory, do not hold in quantum mechanics. This shows that there are some predictions of quantum theory that cannot be obtained from **any** Kolmogorovian theory. This has been put to the test; the work of Aspect, Grangier, Dalibrand and Roger [10, 11, 12] shows that for some systems of two correlated photons, the best experimental estimates for certain spin correlations violate the Bell inequalities, and so cannot be explained by classical probability theory, with or without hidden variables. The statistics observed in the experiments do however agree very well with the quantum predictions. Bohm found a hidden assumption, that of "locality", in Bell's proof of his theorem, but later proofs [110] do not use any such assumptions. In earlier work, Bohm [26] had certainly proposed a theory with a quantum potential which had instantaneous influence over arbitrarily large distances. It is true that modern proofs of Bell's result make the assumption, universal among probabilists since Kolmogorov's book, that there exists a sample space on which all the *non-hidden* variables are random variables. There is no need to make any assumptions about hidden variables, or even whether they are present. A more elaborate version of Bohmian mechanics was suggested by Bohm and Vigier [28], to allow the wave-function itself to be random. But the claim that either of these models gives the same answer as quantum theory to all questions is not correct; we shall see that this claim, made in [28], is refuted by Bell's inequality.

We have seen that, in quantum theory, each state, and set of commuting observables, defines a classical probability theory. We also saw that in this construction, by Gelfand's isomorphism, the sample space depends on which set of commuting variables is being measured. Thus, a given observable is represented by different random variables, depending on what is being measured with it. We say that the representation as random variables is contextual. Fortunately, in quantum theory the probability distribution of an observable is independent of the context. This result does not ensure that there is a unique sample space on which all the observables are random variables; indeed, Bell's inequality shows that this is not possible.

We recall the *EPR* set-up. The Hilbert space of two particles of spin 1/2 is $\mathbf{C}^2 \otimes \mathbf{C}^2$, each factor representing the spin-state of a particle of spin 1/2. Suppose the state is ρ. We have observables $X = 2S_z \otimes I$, $Z = 2I \otimes S_z$ and $W = 2I \otimes S_x$. We see that X commutes with Z, and so they can be simultaneously measured, and the joint frequency distribution $f_{X,Z}(x,z)$ found; the joint probability distribution $p_{X,Z}(x,z)$ can be computed, as the explicit function on the Gelfand sample space $\Omega_{X,Z}$. On a separate collection of samples, we can measure X and W, and get the joint frequency distribution, $f_{X,W}(x,w)$ and the joint probability distribution $p_{X,W}(x,w)$, on the Gelfand sample space $\Omega_{X,W}$. This gives rise to no contradictions. Prekolmogorovians, and some modern statisticians would regard $\left(\Omega_{X,Z}, p_{X,Z}, X, Z\right)$ as one statistical theory, and $\left(\Omega_{X,W}, p_{X,W}, X, W\right)$ as another. Since the joint distribution of Z and W cannot be observed (they are incompatible experiments) we cannot reconstruct, from the data given, a single sample space on which all three observables are random variables; to do this, we would need the joint distribution of Z and W, and also that of X, Z and W. One can reproduce the more limited observed data exactly, even in a quantum experiment, if we allow the choice of statistical model to be *contextual*: it depends on the context, what variable is B measuring?. The need for this freedom of choice is bizarrely called nonlocality by Bohm (it has nothing, or at most, very little, to do with locality in space). In Chap. (9), I offer a conjecture as to how this misnomer arose. We see that the events, (x_i, z_j) are disjoint from the events (x_ℓ, w_k); the marginal frequencies of the variate X could be slightly different, and even (if we are unlucky) the set of values $\{x_i\}$ might differ from the values $\{x_\ell\}$. Since Alice would not know (until later) whether Bob were measuring Z or W, this does not mean that any message has travelled faster than light. Thus, we should expect that the two runs $\{x_i\}$ and the $\{x_\ell\}$ give two runs of samples of the same distribution. This is true in the *EPR* experiment, according to quantum theory. To see this, we need to set up (contextual) probability models, based on theory, not data.

To set up the two models, suppose that $\mathcal{M}$ is a finite dimensional C^*-algebra, containing X, Z, W where X commutes with both Z and W. Consider a quantum mechanical state ρ of the system; this defines states on two abelian algebras, $\mathcal{A}(X, Z)$ and $\mathcal{A}(X, W)$, that generated by X, Z and that generated by X, W, respectively. Then the restriction of the given quantum state to one or other of these subalgebras gives us a state on each algebra, and from this, we get two probability models, by Gelfand's construction: see Chap. (3). Thus, there are two sample spaces, $\Omega_{X,Z}$ and $\Omega_{X,W}$, the sets of characters of the two abelian algebras. Recall that a character of an abelian $*$-algebra is a state obeying the multiplication law (3.23). These sample spaces are furnished with states, [the restriction of ρ to the algebras]; X is represented by the random variable $\omega \mapsto \omega(X)$ on $\Omega_{X,Z}$ in the first case, where $\omega \in \Omega_{X,Z}$ is a character on $\mathcal{A}(X, Z)$. In the second case, X is represented as the random variable $\mu \mapsto \mu(X)$, where $\mu \in \Omega_{X,W}$ is a character on $\mathcal{A}(X,W)$. We see that the random variable used by A to represent X depends on which subalgebra she chooses; even the sample space is different. Care is needed in contemplating this; it is not true that the measurement by B influences A instantaneously. Alice cannot even know what experiment, Z or W, B is performing, until she gets a message (contrary to the claim of R. Penrose, [133]). She cannot set up a classical description of the joint measurement of X and Z (or X and W) until later, when Alice has heard from Bob, as to what he is measuring. Nor does she need to, as she has not yet received any data from Bob. However the distribution of X is the same in both models. This equality is the analogue of Kolmogorov's consistency relations. See Chap. (4).

Similarly, for each ρ and maximal abelian subalgebra of $\mathcal{M}$, we get a statistical model. Gill has pointed out (with tongue in cheek) that this makes the practice of statisticians more general than quantum theory, if we drop the requirement that the models come from some state on $\mathcal{M}$. However, when we add the consistency conditions to each set of commuting projections in $\mathcal{M}$, we reduce to quantum probability. For, we see that the mean of each observable is independent of the context, since (by consistency) the distribution has this property; thus the various statistical models all define the same map from the questions to the interval $[0, 1]$, namely, the mean of the projection. More, since orthogonal projections commute, they all lie in an abelian algebra, which is classical, so the map must be additive on systems of mutually orthogonal projections. We can now apply a theorem of Gleason [71] to show that there must be a state ρ behind the statistical models.

Theorem 1 (Gleason). *Let $\mathcal{M}$ be a finite dimensional $C^{*}-$ algebra and let p be a map from $\mathcal{M}$ to $[0, 1]$ such that, whenever $\{P_i\}$ is a family of mutually orthogonal projections, we have $p(\sum_i P_i) = \sum_i p(P_i)$; then there exists a density matrix ρ such that $p(P_i) = \operatorname{Tr} \rho P_i$ for all i.*

The suggestion of Bohm and Hiley [28] as the solution to the perceived paradoxes in the spin version of *EPR* can be described as prekolmogorovian probability; it adopts the frequentist philosophy, and uses many sample spaces for the same spin observable in different contexts. In this sense, the spin variables are not "elements of reality" in the sense of *EPR*, not even random ones. Later, in [58], it is admitted that momentum is not an observable, as it needs the quantum treatment. The trouble with Bohmian mechanics is that it claims that position is an observable in the Einstein sense, that it is given by a real value for each sample. For this to hold for all time, the difference $x(t_2) - x(t_1)$ must also be a random variable on the same sample space, whereas in quantum theory, this involves p_x, at least in the free theory, and does not commute with $x(t_1)$.

Frequentist probability has no predictive capability, and is merely a noting of the experimental statistics. In Chap. (9), I describe the original (1952) attempt by Bohm at a purely classical interpretation of Schrödinger's equation. I show that the position coordinates of the particle do not give the same statistics as quantum theory.

Bell's theorem, together with the experiments of Aspect et al., shows that the theoretical idea to use hidden classical variables to replace quantum theory is certainly a lost cause, and has been for forty years.

For an account of the *EPR* 'paradox' in the Copenhagen interpretation, see Chap. (6).

8.1 Bell's Theorem

Bohm asked whether the observed statistics, agreeing with experiment, can be obtained from a larger, more complicated classical theory. This is the idea behind the attempts to introduce hidden variables. It is certainly true of the statistics of any fixed complete commuting set of observables; for they form an abelian algebra, and so can be represented by the classical statistics of multiplication operators on a sample space (the spectrum of the algebra). Obviously the full non-abelian algebra cannot be a subalgebra of an abelian algebra, so the way hidden variables are introduced must be more elaborate than extending the algebra by adding them. However, the deep result of J. S. Bell shows (if the dimension of the Hilbert space is 4 or higher) that the full set of statistics predicted by quantum theory cannot be got from *any* underlying classical theory. In the quantum model of two spin-half systems, Bell constructs a sum of four correlations which in a certain state is equal to $2\sqrt{2}$, a factor $\sqrt{2}$ larger than the greatest value allowed in any classical theory.

The original argument of Bell posited the existence of some hidden variables related to the spin measurements of *EPR* by some local equations, in that the spin on the right was not a function of any variables localised on the left, and *vice versa*. Omnès [131] gives an account of Bell's argument. However, Landau has given a proof without any assumption about the existence and properties of hidden variables. Landau's proof uses only that the four spin variables are random variables on a common sample space. In this version, then, we can say that the quantum result cannot be explained by *any* classical probability theory (with or without hidden variables having whatever properties). In this form, Bohm's theory does not escape: it does not give the same answers as quantum mechanics. Omnès allows that it might, because Bell's derivation used locality; in this, Omnès is too polite.

Let us follow [110]. Let P, Q be non-commuting projectors, and also let P', Q' be non-commuting projectors, while P is compatible with P' and with Q', and Q is compatible with P' and Q'. For example, in the *EPR* experiment, P and Q could be projectors onto the value $+1$ of the x and z-components of the spin of the particle on the left, and P' and Q' similar projectors on the right. Define $A = 2P - I$, $B = 2Q - I$, and similarly for A' and B'. For any state ρ define R by

$$R := \rho(AA' + AB' + BB' - BA') = \rho(C)$$

where $C = A(A' + B') + B(B' - A')$. Then $A^2 = B^2 = A'^2 = B'^2 = 1$, so

$$C^2 = 4 + [A, B][A', B'] = 4 + 16[P, Q][P', Q']. \qquad (8.1)$$

Since $\|A\| = \|B\| = \|A'\| = \|B'\| = 1$, it follows that

$$\|[A, B][A', B']\| \leq 4,$$

so $C^2 \leq 8$ and $|R|^2 = |\rho(C)|^2 \leq \rho(C^2) \leq 8$. So in quantum theory, $|R| \leq 2\sqrt{2}$. This result was also found by Tsirelson [164]. If there is a joint probability space on which we can describe $A, \ldots, B'$ by the random variables $f, \ldots g'$ taking the values ± 1, and a measure p on it, then $R = E_p[h]$ where

$$h = f(f' + g') + g(g' - f').$$

Then these random variables commute, so Eq. (8.1) becomes $h^2 = 4$, and

$$|R^2| = E_p[h]^2 \leq E_p[h^2] = 4.$$

So $|R| \leq 2$, (Bell's inequality). Bell showed that the entangled states of the Bohm-EPR set-up give a ρ such that $R = 2\sqrt{2}$, violating this. Thus no description by classical probability is possible.

The famous experiment by Aspect et al. tested Bell's inequalities. This involves observing a system (in a pure entangled state) in a long run of measurements; the correlations singled out by Bell, between several compatible pairs of spin observables, were measured. The experiments showed that R was just less than $2\sqrt{2}$, in agreement with the quantum predictions.

Wightman has said "Anyone who does not think that Bell's theorem is wonderful must have a hole in his head." This does not mean that Wightman thinks that quantum theory requires faster-than-light influences. It means that he thinks that Bell's theorem, together with the photon experiments [10, 11, 12], provides direct experimental evidence that quantum mechanics is correct, and that no classical treatment can be.

Another interpretation of these theoretical and experimental results has worried I. Percival, and led him with Gisin and others to suggest far-reaching alterations to quantum theory. Percival has said that the measuring apparatus is in practice very large, and in the Copenhagen interpretation is a completely classical object, subject to classical laws. And it is such objects that measure the spin-spin correlations and show the results on their dials. But the results are not compatible with classical probability. So, says Percival, something is wrong somewhere. However, although the measuring device is classical, and its springs and wheels obey the classical laws, the *information* shown on the dials is (designed to be) 100% correlated with the properties of the atomic particles. There is no contradiction here; a piece of paper can show a formula representing a law that is not obeyed by the paper on which, or the ink in which, it is written. Dials can show a collection of readings arising from correlations that are not possible for classical objects.

The upshot is that in quantum probability there is no sample space; we have the C^*-algebra $\mathcal{A}$, and this plays the rôle of the space of bounded functions.

Bell's version of proof of his theorem assumed that in addition to the quantum variables there were possibly some others, hidden from us. Some of these variables, collectively denoted λ, were located near the particle on the left, and others, similarly denoted μ, were located near the particle on the right. Bell postulated a locality condition: the spins on the left have no functional dependence on μ, and the spins on the right have no dependence on λ. We have seen that Landau's proof does not need to assume anything about any other random variables than the ones measured, X, Y, Z, W. And these were assumed simply to be random variables on the same space.

It should be said straight away that the solution is to create different sample spaces and observables for each experiment, contrary to the practice of classical probabilists, who would expect there to be a unique random variable representing an observable. It also goes against the definition of "element of reality" of *EPR* as extended by us to the random case. The quantum version does not suffer from this unreality, since the mathematical object assigned to the observable, the Hermitian matrix, does not depend on the context, i.e. is non-contextual; it should thus be called local by Bell and Bohm.

The suggestion in [28], the use of contextual models, might be said to be an interpretation of quantum mechanics in terms of classical probability. However, the construction is not a probability theory in the sense of Kolmogorov, as there is no single sample space; the theory is prekolmogorovian, in the tradition of the frequentist school. One can generalise the frequentist point of view, and specify that certain collections of observations are compatible, and others are not; there need be no addition rule between incompatible elements; then we can by observation construct the joint probabilities of each compatible set, and have no need of an underlying space for them all at once. We can impose a consistency condition, discussed above: if X, Z are compatible, and also X, W, then the distribution of X should be the same in both models. There are no known consistent models apart from those based on quantum mechanics, however.

Various compatible sets need have no analytic relation to each other, even though it contain some common observables. Bell's inequality need not hold, but then neither need the quantum version, which is $\sqrt{2}$ times more generous [164]. This extension is not needed for quantum mechanics; it is a feeble theory, not much more that data collection, and has no predictive power. Mere data give us no more than mere data. An obstacle to constructing more general theories even than quantum probability is Gleason's theorem, as we saw in Chap. (4). Neumaier [127] has proposed a definition of quantum state which reduces to the point $\omega \in \Omega$ for classical systems, and includes the rays of a Hilbert space in a quantum theory. He hopes that, in general, the concept might be able to say, at least weakly, that a system has a state for which non-commuting variables would both have sharp values; he does not solve this problem, though.

9

Bohmian Mechanics

"I have never been able to discover any well-founded
reasons as to why there exists so high a degree of
confidence in the . . . current form of the quantum theory."

D. Bohm, Wholeness and the Implicate Order, p. 84.

Bohmian mechanics was developed by David Bohm [24].
This subject was assessed by the NSF of the USA as follows [Cushing, J. T., review of [28]] ". . . The causal interpretation [of Bohm] is inconsistent with experiments which test Bell's inequalities. Consequently . . . funding . . . a research programme in this area would be unwise". I agree with this recommendation.

The starting point is an idea of Madelung [114], that the quantum probability current has many of the properties of the current of a fluid. To get the full Schrödinger equation of motion, the quantum phase (here, phase means the argument of the complex number ψ) must also be given a dynamics. Bohm found that the phase obeys a sort of Hamilton-Jacobi equation, modified by an extra term involving Planck's constant. He called this extra term the quantum potential. It was hoped that quantum mechanics can be reformulated as a "realistic" theory in the sense of *EPR*. That is, Bohm set up a deterministic dynamics, given by solving the equations; one could introduce randomness of a classical sort by randomising the initial conditions. Then, it was hoped, the randomness of quantum theory would be due to our ignorance of the initial conditions; a similar hope was entertained by some physicists in connection with statistical physics itself. Later, Bohm and Vigier added a stochastic process to the velocity equation, so that the motion was random, given the initial conditions, instead of being deterministic. But the probability is still classical. I shall discuss this version, which was also formulated by [65], in Chap. (11).

Bohmian theory developed and changed with time, and I shall discuss the three books [26, 29, 58] in sequence.

In **Wholeness and the Implicate Order**, Bohm introduces both the deterministic and the stochastic versions of the model, describing the latter as final. The deterministic theory is (p. 76)

1. The wave-function ψ is assumed to represent an objectively real field and not just a mathematical symbol.
2. We suppose that there is, besides the field, a particle represented mathematically by a set of coordinates, which are always well defined and which vary in a definite way.
3. We assume that the velocity of this particle is given by $\mathbf{v} = m^{-1}\boldsymbol{\nabla}\,S$, where m is the mass of the particle and S is a phase function, obtained by writing $\psi = Re^{iS/\hbar}$, with R and S real.
4. We suppose that the particle is acted on not only by the classical potential $V(\mathbf{x})$ but also by an additional 'quantum potential'

$$U = -\frac{\hbar^2}{2m}\frac{\boldsymbol{\nabla}^2 R}{R}.$$

A stochastic version was introduced by Bohm and Vigier [27]. In the stochastic version, [24], p 77, there is a fifth axiom, which is expressed in words:
"... the field ψ is actually in a state of random and chaotic fluctuation, such that the values of ψ used in quantum theory are a kind of average ... in much the same way that the fluctuations in the Brownian motion ... come from a deeper atomic level."
In [28] it is recognised that the velocity equation (3) must be changed by the addition of a stochastic driving term, so that it is changed into a stochastic differential equation. This might be what is meant by axiom (5). The wave-function ψ is expected to be an attractor of the random dynamics thus introduced, and the solution to the Schrodinger equation is hoped to be the mean behaviour of the ψ-field. While this makes the study of the Bohm-Vigier theory harder than the study of Bohm's original model, the new randomness is still classical; there is no difference in the proof that the theory cannot give all the same predictions as quantum theory.

It says (p. 78) "It has been demonstrated [24, 25] that the above theory predicts physical results that are identical with those predicted by the usual interpretation of the quantum theory". This is false. Although written in 1980, there is no mention of Bell's inequality. Instead, there is a discussion of von Neumann's theorem, which was designed to show that hidden variables cannot explain the uncertainty in quantum theory. The theorem showed that there are no dispersion-free states in quantum theory: given any pure state, or wave-function, ψ there exists an observable whose variance is greater than zero. It follows that there cannot be a theory in which all the observables are real numbers in some configuration. Bohm objected to "a certain unnecessarily restrictive assumption behind von Neumann's arguments", which he said (p. 79) is that "... the particles arriving at ... a given position, x, (determined beforehand by the hidden variable) must belong to a subensemble having the same statistical properties as those of an ensemble of particles whose position, x, has actually been measured". Bohm then goes on to say that if the position of the particle is measured, it is "well known" that it spoils the interference patterns. Thus, Bohm uses one of Bohr's arguments, true of quantum theory, to demolish a proof that his [Bohm's] theory does not obey all the rules of quantum theory. To justify the Bohr argument, Bohm relies on the proof that his theory always gives the same results as quantum theory: quantum theory says that a measurement changes the distribution of the position (unless the state is an eigenstate); however, in a classical probability theory, most people assumed that a measurement revealing the value of x but not changing it, is possible. Apparently not if you are Bohm.

The desperate nature of the section on page 79 might have become apparent to Bohmians later, because this argument is not repeated in [28]; instead, Bohm and Hiley are more explicit, and point out that von Neumann assumed that if X and Y are two observables, the mean of their sum is the sum of their means. They say that, if X and Y do not commute, then they cannot be observed simultaneously, and so the relation cannot be tested experimentally, and no meaning can be given to the sum. However, the linearity of the mean is true in quantum probability (which they admit) as well as in classical probability (which they seem to have forgotten). Anyway, it is not true that no meaning can be given to $X + Y$ if X and Y do not commute; in the case of spin 1/2, the sum of two non-commuting spin operators is proportional to the spin operator in some other direction. In the case of the free and interacting Hamiltonians, the sum is the total Hamiltonian. For other operators, quantum mechanics postulates that there is an observable called $X + Y$; it does not always advise the reader how to measure it. In a fully classical theory, there is no problem; we just measure X and then measure Y, (not disturbing the system in either case) and then add them together. Then obviously, the mean of the sum is the sum of the means.

This weakness in the anti-von-Neumann case might have led Bohmians towards the view given in [58], that spin and momentum are not observables, but must be treated quantum mechanically as operators. This view is partially to be found in [28]. Thus, they say, (p 114) "In our interpretation ... what Heisenberg's [uncertainty] principle refers to is not the actual momentum of the particle itself, but the value of the momentum that can be attributed to the particle after ... a measurement of the momentum ... these can differ." They still seem to think that the momentum had a value before measurement. But soon thereafter, (p 120) spins become quantum, as in the passage "there is no *pre-existing* quantity [corresponding to spin values] ... results are not present before the measurement has been completed." These Bohr-like statements are their answer to work beyond the Bell inequality, showing that some quantum results cannot be obtained by any classical theory, even when the quantum results show no uncertainty. Particularly awkward for the authors are two papers of N. D. Mermin [119] and D. M. Greenberger, M. A. Horne, A. Shimony and A. Zeilinger [77]. I shall describe Mermin's neat argument;

the other one is discussed in Chap. (6), as an example of the way to treat counterfactual and hypothetical questions, using consistent histories.

Let $\sigma_{j,x}, \sigma_{j,y}, \sigma_{j,z}$, $j = 1,2$ be the Pauli matrices describing (twice) the spin of two particles of spin $1/2$. Consider the following table of operators.

$$\begin{array}{cccc} \sigma_{1x} & \sigma_{2x} & \sigma_{1x}\sigma_{2x} & 1 \\ \sigma_{2y} & \sigma_{1y} & \sigma_{1y}\sigma_{2y} & 1 \\ \sigma_{1x}\sigma_{2y} & \sigma_{1y}\sigma_{2x} & \sigma_{1z}\sigma_{2z} & 1 \\ 1 & 1 & -1 & \end{array}$$

We can see that the rows of matrices commute, as do the matrices in each columns. So the observables in each row, or column, can be simultaneously measured. The array, then, defines six different experiments that can be done. The product of matrices in each row is the unit operator, and the product of the matrices in the first two columns is also the unit operator, while the product of the matrices in the third column is -1. These are operator identities and so the fourth column and fourth row are sure predictions of quantum theory for the result of measuring a row or column, and multiplying the obtained numbers of the row or column together. (Recall that in quantum probability, the sure random variables are represented by multiples of the identity). This is so whatever the state of the system. No classical probability (or realistic model, in the terminology of *EPR*) can give this prediction. For, if all nine symbols were random variables on a sample space Ω, they would be given by functions of $\omega \in \Omega$. But then, the product of all nine values at ω would be 1, if we multiplying the rows first, and then multiply the answers, but would be -1 if we multiply the columns first. This shows that *there are predictions in any quantum theory whose Hilbert space is of dimension at least four, that cannot be explained by* any *classical theory*. For, in any such space, we can construct operators obeying all the properties of the six Pauli matrices involved. The conclusion is the same as what we get from Bell's inequality.

It is interesting to (try to) follow the argument given by [28] on p 120-121, as to why this no-go theorem does not apply to Bohmian mechanics. The authors start with a false statement: "This contradiction rules out the model of quantum theory that is proposed by Mermin (and of course all the other authors who use essentially the same model)". No! The contradiction is *between the quantum model and any classical model*. It does not say that the quantum model (which is standard, the same for everybody) is ruled out. Experiment shows that every classical model is ruled out. They assert that spin is not a beable (in Bell's sense). They go on: "In our interpretation we do not assign values ... to the operators. ... results are not present before the measurement operation has been completed ... they are only potentialities ... as we have already pointed out, there is no *pre-existing* quantity that is actually revealed in this process." They do not mention that many physicists were pointing this out *to them* for the 30 years between Bell's theorem and 1993, when this was written. The rest of the argument, that Mermin shows nothing, is also familiar. Not all the operators commute; as a consequence of this, we cannot simultaneously measure all nine quantities. Bohm and Hiley do not prove this for all conceivable measuring devices, but rather argue that spin measurements using Stern-Gerlach apparatuses get in each other's way when we try to measure non-commuting spins. This, of course, is what Bohr would have said. The narrative is not very convincing, as shown by the following gem: "... we ought to describe ... how each of the operators above could give rise to the result... With operators such as σ_{1x} and σ_{2y} this can be done fairly easily [using] a Stern-Gerlach process.... However, with product operators like $\sigma_{1x}\sigma_{2y}$ no one has yet suggested any physical process that could lead to results corresponding to their measurements." How about measuring the two commuting σs and multiplying them together? This is just the definition of product, when we represent the operators as random variables on the Gelfand space. Bohmians are very wary of jumping to any conclusions; even in easy cases.

And finally, on p 122, we find "The context-dependence of results of measurements is a further indication of how our interpretation does not imply a simple return to the basic principles of classical physics ... there is no single 'phase space [sample space] for the combined system". So at last, Bohmians admit some of what everyone was telling them since Bohr and von Neumann. However, they still have Bohm's theory to defend, against Bell's inequality and its variants.

The book of Bohm and Hiley has the virtue that it gives a clear definition (p 2): "ontology is concerned with what *is* and only secondarily with how we obtain our knowledge ..."

To summarise my conclusions about [26, 28], they advocate two models, one deterministic and the other stochastic. The only beables are the positions of the particles, the configurations of the measuring apparatuses after a measurement, and the wave function. The latter does not collapse on a measurement. Spin is to be treated as in quantum theory, and while momentum can be measured, the measured value should not be assumed to be related to the velocity of the particles (or to its mean value in the stochastic version).

Neither model is likely to succeed. In the first place, the theory predicts effects that move faster than the speed of light, and this is due to the quantum potential, which is nonlocal. Secondly, the theory without contextuality cannot predict the same results as quantum mechanics, contrary to the claims made by Bohm, and others taking his word for it. What has been shown rigorously [57] is that for any time t, his theory predicts the same distribution for the position X of a particle at time t as does the quantum theory of Schrödinger, and also any function of X is also distributed as in quantum theory. This is achieved by construction, as the equations of motion are designed so that the wave-function obeys Schrödinger's equation. This has led some non-Bohmists to accept the claim that the theory agrees with quantum mechanics in its predictions [131]. While it is then obvious that $|\psi(\mathbf{x},t)|^2$, as a function of t, is the same in Bohmian theory as in quantum mechanics, as claimed by Bohm, no discussion is given of observables at different times. To discuss these questions, we must pass from the Schrödinger picture to the Heisenberg picture. In the deterministic version of Bohmian mechanics, $X(t)$ is a random variable on the initial space of positions, and the only uncertainty is in the initial configuration of the system. The $X(t)$ thus commute with each other for all t, so the probability theory is classical, ie obeys the axioms of Kolmogorov. By Bell's theorem, proved in Chap. (8), the theory predicts that the Bell inequality holds for four correlations between certain pairs of compatible variables. In the quantum theory with a suitable initial wave function, as close to an entangled state as we like, Bell's inequality fails for these four position measurements. Thus, the correlations for these pairs of observables are not predicted to have the same values as in quantum theory. As an example, one might take the four observables, $x(t_1), x(t_2)$ and $y(t_1), y(t_2)$ in the free dynamics. Both $x-$variables commute with both $y-$variables, but if the times are different, the $x-$ variables do not commute, and neither do the $y-$variables. To get exactly the Bell set-up, we should consider projections onto an interval, say onto small intervals around ± 1 in all cases. Then we seek an initial state such that Bell's inequalities are violated in the quantum theory. It could be the analogue of the entangled state that worked in Bell's original paper. We suggest that such a state can be at least approximately constructed experimentally by splitting a single slow neutron by half-reflecting mirrors, passing the two halves through various slits in the $x-$ and $y-$directions, and bringing the beams together. With this preparation, we measure the four correlations that enter the Bell inequalities, using the distance along the direction of motion of the beam as the time variable. Quantum theory then predicts that the Bell correlations sum up to something bigger than 2.

We now turn to the latest and most mathematical of the books on Bohmian mechanics [58]. It is argued there, confirming Bohm's idea, that all observables ultimately come down to position measurements of the particle. It is not clear, however, whether the supporters of the theory expect all the position measurements needed to measure say momentum can be done at the same time. As Fine remarks in the same book, page 235, a classical theory predicts the form of the joint distribution of observables that do not commute in the quantum theory, and so have no joint distribution in quantum mechanics. In quantum theory, the positions of the particle at different times do not commute, whereas Bohm's theory predicts their joint distribution. This is an indication that there are observables whose distribution is not correctly given by Bohm's theory. Indeed there are. This easy fact, an immediate consequence of Bell's inequality, is denied by most of the authors in [58].

The classical velocity of the Bohmian particle is not related to the measured values of the velocity in quantum theory. In any real ψ, the Bohmian velocity is zero, whereas the measurements (as predicted by quantum theory) is a random variable with distribution given by $|\phi|^2$, where ϕ is the Fourier transform of ψ. The ground state of the hydrogen atom is real for all time, and so in Bohmian physics, nothing moves. This is not a contradiction, according to Fine (p. 245): "... reality ... is significantly different when observed than when unobserved". Squires (p 133) uses the equality of the quantum and Bohmian distributions of position (which does not, in fact hold for the joint distributions of positions at two different times) to argue that the incorrect Bohmian value of the velocity is no problem. He claims that we can measure the momentum by measuring the positions of certain probes. These "will be correct". However, one can imagine measuring the momentum of a particle by measuring its position at two different times, (and dividing the distance by the time interval); in Bohmian theory this will give a measure of the average Bohmian velocity over that time interval. Squires is saying that he does not believe that this velocity is what will be observed in any experiment. So much for the reality of the model.

The theory is not consistent in its interpretation. It is set up as a probability theory; but a measurement of the position of the particle does not result in a conditioning of the probability distribution of the position. This is at the insistence of J. S. Bell, who championed the Bohm theory when it was more or less discredited. Bell said "No-one can understand this theory until he is willing to see ψ as a real objective field rather than just a probability amplitude" [20] p. 128. This allows believers to claim that there is no measurement problem, as a measurement just reveals what the position *is*. The usual rules of probability would lead to the replacement of $|\psi|^2$ by a conditional probability, after a measurement has given us more info as to the whereabouts of the particle. However, this conditioning is exactly the collapse of the wave packet, and this is a dirty word among Bohmists: to allow it here might weaken the case against quantum mechanics.

The failure to condition after measurement leads to peculiar results even for positions. Y. Aharonov and L. Vaidman [28] (pp 141-154) in "About position measurements", give many examples of this, similar to those in [61]: Aharonov and Vaidman admit: "We worked hard, but in vain, searching for an error in our and Englert [*sic*] et al arguments". They conclude "The proponents of the Bohm theory do not see the phenomena we describe here as difficulties of the theory", and quote a riposte of the Bohmists [56].

After the results of the experiment by Aspect et al. were known to be inconsistent with a classical model with hidden variables, Bohm called a press conference. He said that "contrary to the expectations of most physicists, the experiments show a violation of Bell's inequalities". On the contrary, the experiments, agreeing with the quantum predictions, were expected by the vast majority of physicists. Squires (*q.v.*) expresses a more up-to-date view of the Bohmists: "Any 'completion' of quantum theory which is consistent with all its predictions must be nonlocal .... most physicists have come to accept this nonlocality in some form or other". WRONG! most physicists accept the Copenhagen interpretation, in which quantum probability does not obey Einstein's concept of reality, but is both local and non-contextual. This is our conclusion in Chap. (4).

Einstein has given the following definition: In any theory, a concept is called an *element of reality* if is assigned a [real] value. He goes on to say that if an attribute can be measured without in any way altering the system, then it should be represented by an element of reality [in any good theory]. The famous *EPR* experiment, in which the momentum of a particle is 100% anti-correlated with another at a far distance, shows that by measuring the momentum of the far particle we can find that of the first. Naturally, *EPR* thought that this does not change the first particle in any way. So they concluded that momentum should be classed as an element of reality [in any good theory]. Bohm concocted a similar *EPR* pair, in which the spins are 100% anti-correlated. Thus we might conclude that Bohmists regard both the momentum of a particle, and its spin, as elements of reality. Not a bit of it! According to [58], pp 21-44, spin, momentum, and energy, in Bohmian mechanics are not elements of reality; they must be considered as self-adjoint operators, whose measured values are given by their eigenvalues, and whose probability distribution is given by the expectation value of positive-operator-valued measures in the state ψ. But this is quantum probability, as formulated by E. B. Davies and J. T. Lewis in [50], and expounded in the book [48]. To make sure the reader does not use it for position measurements as well, it is denigrated in [58] as a mere technicality. It is "even more abstract than operators", (p 32), but is computed "by elementary functional analysis ... [and] is a near mathematical triviality". Nevertheless, it was used in [58] for the theory of spin measurements when explaining the violation of the Bell inequalities as found experimentally in [10, 11, 12]. This was done because they do not regard as real enough any symbol whose meaning must be changed when the context is changed, and this contextuality is needed if we try to explain the Aspect experiment by classical probability, as in Chap. (8).

The use of Davies-Lewis theory for spin and momentum in [58] is bravado. This theory is a generalisation of von Neumann's theory, to include the measurement of parameters like time which is not represented by an operator in von Neumann's version. This generalisation is not needed here, and one can make do with the usual (von Neumann) projector-valued measures for the spin and momentum.

Einstein's definition of "element of reality" needs elaboration to meet modern standards of mathematics. Since a system can be in various configurations or states, the real number assigned to the element of reality cannot always be the same. So the definition should have said, "In any theory, a concept is called an *element of reality* if, in a given configuration of the system, it is assigned a number." With this definition, we may call $\Gamma = $ *phase space*, the set of pure states, and then an element of reality is a map from Γ to the reals. If we add the technical requirement of measurability, this is what is called an observable in modern classical dynamics.

The definition can be adapted to a theory with randomness, given by a measure μ on Γ: "In any theory with randomness, a concept is called an *element of reality* if it is given by a random variable X on Γ." This is none other than an element of reality as above: it is assigned a real value for each $\omega \in \Gamma$. Although Einstein was against randomness ("HE does not play dice"), he used it a lot in some of his best papers. Bohmists take this as the definition of reality; see [58], bottom of p 35. For them, energy, momentum and spin are not elements of reality. However, the claim that position is an element of reality (in Einstein's sense) at all times, true of Bohmian mechanics, is not true in quantum mechanics, contrary to their belief. In quantum mechanics, positions at different times do not commute, so there is no representation of them by random variables on a common sample space: some correlations referring to positions at different times, fail to satisfy Bell's inequalities, which they would do if there were such a representation. I have given examples of these, and a suggested experiment, to test Bell's inequalities arising from positions, in Chap. (8).

We do not wish to define the concept, element of reality, as above in the manner of Einstein; for then in quantum mechanics, there are no elements of reality. We can generalise the concept from classical to quantum probability as follows, bearing in mind that self-adjoint operators are the quantum analogues of random variable: "In any quantum theory, based on the Hilbert space $\mathcal{H}$, a concept is called an *element of reality* if it is a self-adjoint operator X on $\mathcal{H}$. A state is a density operator ρ on $\mathcal{H}$, and the mean of any element of reality in a state ρ is Trace(ρX)." We see here the von Neumann definition of observables in quantum mechanics; it must be generalised if there are superselection rules.

Like the Bohmists, [58], I do not wish to call "elements of reality" any assignment (of operators to concepts) which must be changed to another assignment if the context is different. As an example of a contextual assignment, any model agreeing with experiment, and which describes spin as a random variable, must use different sample spaces depending on the context. On the contrary, contextuality is not needed in the quantum theory of spin. This point has been made before, and I repeat the argument: suppose we have the *EPR* pair of spins $1/2$ particles as in Bohm's modification of *EPR*. Take $\mathcal{H} = \mathbf{C}^2 \otimes \mathbf{C}^2$, where $\mathbf{C}^2$ is the two-dimensional complex Hilbert space. We may assign the operators $\mathbf{s} \otimes I$ to the spin of a particle on the left, and $I \otimes \mathbf{s}$ to the spin on the right, where $2\mathbf{s}$ is the vector of Pauli matrices. Then these two operators commute elementwise, so the measurement of one spin does not disturb the other (locality). Moreover, we do not need to change $\mathbf{s} \otimes I$ to various other operators depending on what simultaneous measurement is being done to the other particle, in order to agree with the Aspect experiment. Thus the assignment is non-contextual. We conclude that quantum probability allows a realistic, local (= non-contextual) description of the Aspect experiments, provided that we modernise Einstein's concept of reality. This is the view of the vast majority of physicists, contrary to Squires's statement.

Bohm's claim that his theory gives the same results as quantum theory has been accepted by some physicists, including Stapp [152]. Why not test Bell's inequality on an entangled state in Schrödinger theory, to show that Bell is violated, while Bohm's theory obeys Bell? This shows that the two theories make different predictions. Stapp gets out of this one by agreeing with Bohm that Bell made a hidden assumption in his proof of the inequality, namely, that the theory was local. Stapp knows that Bohm's model has a nonlocal term in the equations of motion, namely the quantum potential, and assumes that it is this term which changes the value of an observable at one point instantly another is measured, so as to cause larger-than-life correlations. Stapp gives no detailed calculation, as to how this works. No need of this; he says it must be so; how else could Bohm's theory reproduce all the results of quantum mechanics?

In Landau's proof of Bell's inequality, given in Chap. (8), there is no mention of the dynamics at all; we look at the correlations between four observables at the same time (in a series of runs with the same initial state). In [58], the quantum results are got by introducing quantum operators for the energy, momentum and spin; Bohmian dynamics is only used for the position variable, and even then no agreement with quantum theory is shown for position measurements at different times. No such agreement is possible: in Bohm, all position observables (at all times) are random variables, and so any four observables constructed out of them, and having values ± 1, would obey Bell's inequalities. In Schrödinger theory, however, there will be some commuting pairs of non-commuting operators, functions of the positions at various times, that violate the Bell inequalities. These can be measured in pairs on a run of samples. Thus even the latest version of Bohmian mechanics [58] will not give the same results as quantum theory. In recent preprints [126, 69], Neumaier and Ghose have some examples. Neumaier considers the measurement of $X(t)$ for a single harmonic oscillator at two times differing by half the period. He finds that $\mathbf{E}[X(t_1)X(t_2)]$ given by Bohm's theory is the negative (not zero) of that given in quantum theory.

More recently, Goldstein still maintains that Bohmian mechanics gives the same results as quantum mechanics. However, this statement requires that the measurement theory is brought in line with the quantum rules. Thus, when the wave function is ψ, and we measure the position of the particle at time $t = 0$, so that $a \leq X(0) \leq b$ is found to be true, then we must alter the wave function at time 0 from $\psi(x)$ to

$$\phi := \frac{\chi_{[a,b]}(x)\psi(x)}{\left(\int_a^b |\psi(x)|^2 \, dx\right)^{1/2}}. \tag{9.1}$$

The propagation of this initial value to the solution at any time $t > 0$ by the equation of Schrödinger, or by Bohm's equations, then both give the same result; and $|\phi(x,t)|^2$ gives the probability density that the position is at the point x, as it should. This modification of ψ to ϕ at time zero is not, however, what one would do in the classical stochastic process $X(t)$ when it value is found to lie in $[a, b]$.

Better steer clear of Bohmians.

10

The Analytic S-matrix Bootstrap

It may be flogging a dead horse, but I will mention this subject, and why it failed.

10.1 Scattering

The S-matrix is the operator that, in a given theory, transforms free, ingoing beams of particles into free, outgoing beams of particles. This operator can be proved to exist in the scattering of a particle from a scattering centre, represented by a potential of short range, and in the theory of mutual scattering of non-relativistic particles with a mutual potential of short range. It can also be proved to exist in any relativistic quantum field theory, if we know that all particles are massive; this is one of the well-known triumphs of Wightman's version of local quantum field theory, and is known as the Haag-Ruelle theory. The first case, scattering theory using the Schrödinger equation with a fixed external potential, is taken as a good approximation to the scattering of a particle of small mass from another of large mass, as in the scattering of an electron from a nucleus. The theory is not invariant under the Galilean group, and is regarded as a toy model of the second or third case. While energy is conserved in this dynamics, momentum is not, and no account is taken of the recoil of the heavy target. It is, nevertheless, a good theory to start with, when considering the properties of the S−matrix. It turns out that similar properties, suitably formulated, hold for few-body scattering using the few-body Schrödinger equation (invariant under the Galilean group) and in Wightman theory, which is invariant under the inhomogeneous Lorentz group.

In the quantum theory of a single particle of mass m moving in a potential V, invariant under the rotation group, the energy operator is

$$H = \frac{p^2}{2m} + V(|\mathbf{x}|) = H_0 + V. \tag{10.1}$$

acting on the domain of states of finite energy lying in the Hilbert space $\mathcal{H} = L^2(\mathbf{R}^3)$ of square-integrable wave-functions. Then the Schrödinger equation

$$i\hbar \frac{\partial \psi}{\partial t} = H\psi \tag{10.2}$$

can be formally solved, giving the time-evolution operator $U(t)$ in terms of the exponential of the Hamiltonian:

$$\psi(t) = U(t)\psi(0) = e^{-iHt/\hbar}\psi(0) = (1 - iHt/\hbar - H^2t^2/\hbar^2 + \ldots)\psi(0). \tag{10.3}$$

The free Hamiltonian H_0 can also be exponentiated to give the free time-evolution $U_0(t)$. At large positive or negative times the wave is a large way from the scattering centre, and we might expect the particle to behave as a free particle. This can indeed be proved, in the sense that $U(t)^{-1}U_0(t)$ converges to a constant operator $\Omega_\pm$ as $t \to \pm\infty$; $\Omega_\pm$ are called the wave operators, and are unitary if there are no bound states. Notice that $U(t)^{-1}$ and $U_0(t)$ do not exactly cancel each other out at infinite time, but converge to a unitary operator representing the scattering caused by V, which bends the path of a wave packet from a nearly constant direction at ∞ to a different one at time $t = 0$. The scattering operator S is then $S = \Omega_-^* \Omega_+$. One proves that S is a unitary operator for a huge class of potentials.

This, the modern form of scattering theory, replaces some earlier methods. For example, there is the method of "adiabatic switching"; one assumes that the interaction potential becomes zero exponentially fast at large and small times, by replacing V by $e^{-\alpha|t|}V$, where $\alpha > 0$ is made to converge to zero after the calculation is over. In fact, for potentials V of short range, the potential automatically becomes small as the particles move apart, and this approximation is omitted in modern accounts.

To explain the method used in the S−matrix bootstrap, and why it cannot mean anything, we stay with the class of rotation-invariant potentials, which leads to properties similar to those expected to hold in the relativistic case. First, the energy is conserved in a scattering, and, because of rotation invariance, the angular momentum is conserved. The ingoing plane wave can be written as a superposition of waves of given energy and angular momentum. Energy runs from 0 to ∞, whereas angular momentum is quantised. The space of ingoing states is spanned by wave-functions of given energy, given values $\ell(\ell + 1)$ of $J_x^2 + J_y^2 + J_z^2$, and $J_z = -\ell, -(\ell - 1), \ldots, +\ell$. Regarded as an operator in a suitable sense, the S−matrix must conserve these quantum numbers, and so cannot do more than change the eigenstates of these operators by a phase. This is because here, we have a complete set of commuting operators with eigenvalues (E, J, J_z), and the corresponding Hilbert space is of dimension one. That V is rotation invariant leads to the result that S is independent of the azimuthal quantum number, and is a function only of energy and scattering angle. When expanded in Legendre polynomials, S is then determined by the phase shifts $\theta_\ell(E)$, $\ell = 0, 1, 2 \ldots$; $e^{i\theta_\ell(E)}$ are called the partial-wave amplitudes: S-wave for $\ell = 0$, P-wave for $\ell = 1$, D-wave, for $\ell = 2$, ...

It is clear that unitarity and analyticity do not determine the form that V must take, since both these properties follow for a huge class of different choices of V. So we might try to add special relativity and causality as more constraints, to see whether these do fix things.

10.2 Dispersion Relations

The dispersion relations we shall deal with are a consequence of the fact that $S(E)$ is an analytic function of the variable E, and so the value of $S(E)$ at any point is determined by the values on the boundary of the domain, by Cauchy's integral formula.

The idea of causality is that a signal cannot be observed before it has been sent. This idea should hold in non-relativistic theories (if they are any good) as well as relativistic theories. In the latter, the principle must be strengthened, to prevent signals from travelling faster than light. If a wave-packet does not arrive in the locality of the scattering centre before time $t = 0$, then the outgoing wave must be zero for negative times. In quantum mechanics, the wave-function $\psi(t)$ is the Fourier transform of the wave-function $\hat{\psi}$ regarded as a function of energy. Thus we may write

$$\hat{\psi}(E) = (2\pi)^{3/2} \int_0^\infty e^{iEt} \psi(t) dt$$

Under mild conditions on ψ, the integral converges also for *complex* values of E, provided the imaginary part is positive. It defines an analytic (= holomorphic) function of energy in the upper half-plane. If this analytic function, $\hat{\psi}(E)$, is polynomially bounded, then there is a converse, which says that $\psi(t)$ is zero for negative times. Now, S converts the incoming wave into an outgoing wave, which must also be zero for negative times. But each partial wave is just multiplied by $S_\ell(E)$, so the product $S_\ell(E)\hat{\psi}(E)$ must be analytic. This more or less forces $S_\ell(E)$ to be analytic in the upper half-plane, and polynomially bounded. It turns out that in scattering theory, this can be proved to be true both for scattering off a scattering centre, and for the two-body scattering theory where the target is a recoiling particle.

For the solution of quantum field theory, the function $S_\ell(E)$ should be a function which is analytic in the upper and lower half-planes, but with a cut on both the positive and the negative real axes; there is a gap in the cut around zero, provided that there are no particles of zero mass in the theory. Thus there is a cut from $-\infty$ to $-E_1$, which is negative; and another cut from $E_2 > 0$ to $+\infty$. We call the numbers E_1, E_2 the *thresholds*. The cut for negative energies comes from the antiparticles that are present in a relativistic theory. This fact can be proved for models that obey the Wightman axioms, provided that the variable, the momentum transfer, is not too large [33]. Then, Cauchy's theorem on contour integrals tells us that

$$S(E) = \frac{1}{2\pi i} \int_{\mathcal{C}} \frac{S(z)}{z - E} dz, \tag{10.4}$$

where $\mathcal{C}$ is any closed contour inside the region where $S(z)$ is analytic, and E is any possibly complex value inside the contour $\mathcal{C}$. The contour integral, Eq. (10.4), is defined in terms of the equation for $\mathcal{C}$. Suppose that

$$\mathcal{C} = \{z \in \mathbf{C} : z = \xi(t) + i\eta(t), 0 \leq t \leq 1\} \tag{10.5}$$

where $\xi(t)$ and $\eta(t)$ are given smooth functions, and are periodic with period 1. The value to be integrated along the contour is then a function of the position $(\xi(t), \eta(t))$ on the contour; more, the variable $z = \xi(t) + i\eta(t)$, and so

$$dz = d\xi(t) + i\, d\eta(t) = (\dot{\xi}(t) + i\, \dot{\eta}(t))dt.$$

Consider a complex valued function, $f(x+iy) = u(x,y) + iv(x,y)$ defined in a complex region. Then the contour integral of f along the contour $\mathcal{C}$ is defined in terms of real integrals as

$$\int_{\mathcal{C}} f(z)dz = \int_0^1 dt\, (u(\xi(t), \eta(t)) + iv(\xi(t), \eta(t))) \left(\frac{d\xi}{dt} + i \frac{d\eta}{dt} \right).$$

Thus, in the contour integral, Eq. (10.4), we put $f(z) = \frac{1}{2\pi i} \frac{S(z)}{z-E}$, which is assumed to be analytic in a neighbourhood of the contour $\mathcal{C}$.

In the case in point, a *dispersion relation* in scattering theory chooses $\mathcal{C}$ to be wrapped around the two cuts in the energy plane; it comes in from $-\infty$ just above the cut, passes through the gap at $E = -E_1$ and then goes back to $-\infty$ just below the cut. We denote the small distance above and below the cut by ϵ. The contour also comes in from $+\infty$ just below the cut on the positive axis, and goes through the gap at E_1, and returns to $+\infty$ just above the cut. $\mathcal{C}$ is completed by adding large semi-circles in the upper- and lower half-planes. If we assume that the function falls off fast enough at ∞, we then can convert Eq. (10.4) into the *dispersion relation*

$$S(E) = \frac{1}{2\pi i} \left\{ \int_{-\infty}^{-E_1} \left(\frac{S(x + i\epsilon) - S(x - i\epsilon)}{x - E} \right) \right. $$
$$\left. + \int_{E_1}^{\infty} \left(\frac{S(x + i\epsilon) - S(x - i\epsilon)}{X - E} \right) \right\}. \tag{10.6}$$

In Eq. (10.6) we have taken the contribution from integrating round the large semi-circles to be zero in the limit, as S is taken to be small at infinity. When we take $\epsilon \to 0$ we do not get zero along the cuts, since the function $S(E)$ is discontinuous across the cuts. It can be shown that this discontinuity is related to the square of the scattering amplitude, assuming the theory is unitary. Thus, the dispersion relation gives us some nonlinear relations between these amplitudes.

More is true. The partial-wave expansion

$$S(E, \theta) = \sum_\ell S_\ell(E) P_\ell(\theta)$$

should converge. Thus we get that the amplitude $S(E, \theta)$ is holomorphic in the two variables, energy and scattering angle.

10.3 Mandelstam's Formula

Mandelstam took this idea and extended the domain to be the maximum possible, which he guessed to be the double complex plane except for six cuts, two, over positive and negative values, in each of the variables

$$s = (p_1 + p_2)^2, \qquad t = (p_1 + p_3)^2, \qquad u = (p_1 + p_4)^2.$$

In these equations, the square of the four-vector p_i is the relativistic square:

$$p^2 := p_0^2 - p_1^2 - p_2^2 - p_3^2.$$

These variables are related by

$$s + t + u = m_1^2 + m_2^2 + m_3^2 + m_4^2$$

where $m_1, \ldots, m_4$ are the masses of the four particles:

$$p_i^2 = m_i^2.$$

Thus, the function S is a function of two complex variables, s, t (or s, u or t, u; the three variables s, t, u are linearly related to each other). In each pair say s, t there are two cuts, for positive and negative values, of s and also for t. Similarly, there are two cuts in each of the other two pairs, s, u and t, u. When viewed in the s, u variables, or the t, u variables, the physical values of S refer to the scattering amplitudes of a particle against an antiparticle, or of the scattering of two antiparticles. All these processes are related to each other in a local quantum field theory.

Now apply the Cauchy contour integral twice, around the cuts, to get

$$S(s, t) = \left(\frac{i}{2\pi}\right)^2 \left\{ \int_{-\infty}^{-M_1} + \int_{M_1}^{\infty} \right\} ds' \left\{ \int_{-\infty}^{-M_2} + \int_{M_2}^{\infty} \right\} dt' \frac{K(s', t')}{(s - s')(t - t')}$$

+ similar expressions with t, t' or s, s' replaced by u, u'.

Here, we have replaced the contour at infinity by zero, assuming that S vanishes there. The function K is given by

$$K(s', t') := S(s' + i\epsilon, t' + i\epsilon) - S(s' - i\epsilon, t' - i\epsilon)$$

for small ϵ; this must obviously be changed suitably for the two other terms.

If K does not fall off at ∞ fast enough for the contour integral described to converge, we subtract its supposed polynomial behaviour at ∞, and this introduces new parameters into the theory. We then write the dispersion integral for the subtracted quantity. This is the Mandelstam representation. The number of new quantities introduced by the method is the analogue of the number of CCD-zeros (after Castilleijo, Dalitz and Dyson) in the functions.

After the success of Mandelstam's double dispersion relations in 1958, Chew and Mandelstam applied it to the pion-nucleon system, in a truncated form. The S-matrix must be unitary. The elastic region is the region of two-particle scattering at energies before there is enough to produce new particles. Here, the unitarity of the S-matrix amounts to the parametrisation of the scattering by the partial wave scattering lengths, one for each eigenstate of the isotopic-spin group; we suppose that the $P - N$ system are the two states of an iso-spin 1/2 doublet, and the three pions form an iso-spin three-vector; the interaction between them is then supposed to be invariant under the iso-spin group. This implies that the $S-$matrix commutes with the iso-spin group, and so has the same values on the eigenstates of the total isospin. Suppose we include only the four channels, having angular momentum 1/2 or 3/2, and isotopic spin 1/2 or 3/2. This is described by four scattering lengths, each a function of energy. Each of these obeys a dispersion relation, and so can be expressed in terms of an integral of the corresponding imaginary part over all energies. The imaginary part is related by unitarity to the unknown scattering lengths themselves. Thus four nonlinear integral equations in four unknown functions are obtained. The numerical solution of these equations gave the very welcome prediction that there should be a resonance in the (3/2,3/2) channel. That is, states with angular momentum 3/2, and isotopic spin also 3/2, should have a very large crossection, and the predicted energy of this resonance was near to the experimentally observed (3/2,3/2) resonance. The only inputs needed were the pi/proton mass ratio, m say, and the strength, g, of the strong coupling constant; of course, we also assume that the interactions are invariant under isotopic spin and Lorentz transformations.

10.4 Bootstrap

Dashen and Frautschi saw their way to do better, and their idea can be schematically explained as follows. We know that if we have N unknowns, and they are to obey N nonlinear coupled equations, then typically, we expect there to be a discrete set of solutions, many of which might be ruled out for physical reasons. We might hope that this would lead to a unique solution. Indeed, there is only one world, they argued, so the correct set of equations must lead to it in a unique manner. This is, I should add, an unsound argument, unless we add that the equations we use are complete as well as correct. And completeness can have no other general definition other than that the equations have a unique solution. It follows that we cannot use the principle of uniqueness to find the very equations themselves; for this would tell us (if we omitted a true but needed condition) to reject as incorrect a partial system of equations which is correct. It was noticed that in S-matrix theory we have infinitely many equations, expressing unitarity of S, for infinitely many unknowns, the scattering lengths at each energy.

In the bootstrap method, one picks oneself up by one's own bootlaces as follows: regard the pi/proton mass-ratio m and the coupling constant g as unknowns, as well as the four scattering lengths at all energies; we keep the same equations as were successfully used to predict the (3,3) resonance. We still have the same number of equations as unknowns (infinite), and so there might be a unique solution, or even better, a discrete spectrum of solutions predicting all masses of the baryon resonances. We truncate the infinite set of equations to a finite set, with unknowns listed as m, g, and a finite set of N phase-shifts in a finite range of energy. Together, these make up the unknown vector X. We then keep $N + 2$ of the unitarity equations, which express the vector X of unknowns as a nonlinear function F of its components:

$$X = F(X)$$

To solve, we use the iteration method: guess an initial value for the unknowns, such as $X(1) = (m(1), g(1), 0, 0...)$; this is called the input. We put it in the nonlinear term, thus getting the output of the first iteration:

$$X(2) = F(X(1))$$

Continue, until we get convergence. Unfortunately, the method did not converge very rapidly, if at all. So a better method is to consider the difference,

$$X - F(X)$$

as a nonlinear expression in the components of X, and find the conditions that its square be minimised, by varying X. Then at least the error (in making the original truncation, perhaps) would be small. This gave a prediction for all the constants m, g etc of the system. Unfortunately, when more energies were included, or three-particle corrections were made, the results were not close to the previous approximation. When we first heard of this at Imperial College, Paul Matthews said: the bootstrap method consists in making a poor approximation to an exact algebraic identity, and then asking for what values of the fundamental constants make this approximations as least bad as possible. We had already met this idea in a primitive form, in the (wrong) work of one of our students. This work can be summarised schematically as follows. We know that the phase of the $S-$matrix in a given channel can be written as $e^{i\theta}$. This was commonly approximated by $1 + i\theta$ in the literature, where the obvious condition, small θ, was not always brought to the student's attention. Our student saw the opportunity to apply the requirements of self-consistency, and wrote

$$e^{i\theta} = 1 + i\theta$$

which leads to $\cos\theta = 1$ and $\sin\theta = \theta$. He rejected the only solution, $\theta = 0$, as representing the free particle case. His conclusion was that scattering theory is inconsistent with interaction!

Anyone who has worked in quantum field theory would have expected that there is more than one theory: in quantum electrodynamics we can put in any value for the fine-structure constant α, instead of the actual value, and do the same calculations. We would not expect that general principles alone would determine α. In non-relativistic scattering theory, a large number of different potentials lead to a unitary and analytic $S-$matrix. Thus, we might have thought that the equations of the bootstrap method are algebraic identities for each choice of m, g. What happens when we approximate the identities, and then solve for the parameters, including m and g? We get a numerically unstable system which gets worse the closer we approximate the algebraic identities.

My fellow student, Claud Lovelace, became a Senior Fellow at CERN, and he proved that the bootstrap equations had no solution. When Euan Squires gave us a seminar at Imperial College on his research during his stay on the West coast of USA, I asked him whether he was worried by Claud's result. He looked at me rather shocked, but saw a way out: "Aha, you mean that the equations have no *exact* solution. We are physicists, and are looking for *approximate* solutions!" For non-experts, I should say that there is no such thing as an approximate solution to an equation with no solutions; the best you can hope for is a solution of an equation that approximates the terms of the recalcitrant equation. As such, it is not a unique concept, since any equation, or system of equations, can be regarded as the small α limit of any number of quite arbitrary changes to the equations, where the changes are proportional to α. In the case of the bootstrap equations, the dispersion integral probably diverges, and needs a subtraction, an adjustment that introduces new unknowns into the equations. Then one cannot predict all the constants of nature from them.

I wrote out a refutation of the bootstrap method, and submitted it to Physical Review Letters; but when the referees complained that it was not comprehensible, I withdrew it, rather than rewrite it; I was not an expert of numerical instability.

Let us show, by a simple example, the unwisdom in the bootstrap philosophy. Consider the good system of equations for the unknown $X = (x_1, \ldots)$:

$$x_1 = b$$
$$x_2 = 2^{-1}(1 + x_1)$$
$$x_3 = 2^{-2}(1 + x_2)$$
$$\ldots$$
$$x_n = 2^{-n+1}(1 + x_{n-1})$$
$$\ldots$$

We can solve in sequence, and if b is real we get the solution

$$x_2 = (1 + b)/2; \qquad x_3 = 2^{-2}(1.5 + b/2); \qquad x_4 = 2^{-3}(1.375 + b/8); \ldots$$

One easily proves that as $n \to \infty$, then $x_n \to 0$, whatever the value of b. If $b > -1$ we see that all solutions are positive, and go to zero faster than any power of n, as $n \to \infty$.

Let us now try to solve for b as well as the x_n. We will make a small error if we put all $x_n = 0$ for $n > n_0$. There are then $n_0 + 1$ unknowns, so we must keep the first $n_0 + 1$ equations. The $(n_0 + 1)^{\text{th}}$ equation then says that

$$0 = 2^{-n_0}(1 + x_{n_0})$$

which leads to $x_{n_0} = -1$. From this we find that all other xs are large and negative, and go to minus infinity as $n_0 \to \infty$. Thus, we see that there is no approximate solution to the equations, at least among those that converge as $n_0 \to \infty$. This is true, even though the claim made above, that we will make a small error if we put $x_n = 0$ for all $n > N_0$ is true. This holds in that $x_n \to 0$ as $n \to \infty$; however, it still led to an unstable system of equations.

Immediately after the paper of Dashen and Frautschi, it was refuted by several authors [104, 107, 105, 132, 143, 15] . The situation is explained by Kim [104]) which was half-explained by Dashen and Frautschi in a later paper [45]. They say

"In our argument, the requirement of self-consistency for the strong interactions will take the following relatively non-controversial form. Given some physical quantity x_i, such as a coupling constant or the mass of a particle, we suppose that x_i can be calculated in terms of other physical quantities x_j and possibly some additional parameters C ... .[For] C we have in mind parameters such as those associated with a CDD pole in a dispersion relation or the bare masses and couplings of an elementary particle in the language of field theory

$$x_i = f(x_1, x_2, \ldots; C).$$

Of course, in a true bootstrap theory, there are no 'CDD-pole'-parameters C, but we are considering the more general situation". It is not clearly more general, but is more stable. They do admit that a bootstrap containing a single

octet of vector mesons always ends up with some undetermined parameters-
"one cannot tell if a solution [without parameters] really exists." It is clear
that it does not.

Indirectly, the bootstrap theory led to the writing of my book, **PCT, Spin
and Statistics, and All That**, with Arthur Wightman [159]. I think it was
1961, and Wightman went as usual to the "Rochester" conference on particle physics. In the then recent past, people like him and Lehmann, Källén,
Symanzik, Zimmermann and Res Jost had been welcome at the meeting, to
explain the new rigorous results. That year, their slot was occupied by the new
bootstrap analytic S-matrix theory. After the talk by Geoffrey Chew, Wightman asked him why he *postulated* analyticity, rather than founding it on the
axiom of causality. Chew said, all physical functions are analytic, because the
functions of physics are always smooth. Then Wightman asked, what about
infinitely differentiable functions that were non-analytic? Chew thought for
less than ten seconds, and then said "There aren't any". This incorrect as-
sertion so annoyed Wightman that he complained to Bethe, that the rigorous
people had been excluded to make way for someone who did not know what an
analytic function was. Bethe replied that the rigorous theory was out of reach
to most of the audience, and that we, the people concerned, should explain
its most important results in a short book. So this is what we did.

Nelson and Wiener

"Unperformed experiments have no results"
A. Peres.

After the efforts of Fényes and Bohm to explain the quantum uncertainty classically, two great mathematicians, Wiener, and Nelson, took up the challenge. Wiener collaborated with Amand Siegel in three papers, [168, 169, 148], written before Bell's theorem. Nelson published his attempt in two books [122, 123].

11.1 Wiener

In the book by Wiener, Siegel, Rankin and Martin [170], Bohmian dynamics is said to agree with quantum mechanics in all predictions. Wiener had died, and the other authors wanted to publish more widely the theory of Wiener and Siegel of 1956. Note that this was well before Bell's theorem, which started to happen in 1963. The upshot of [170] was to have been a proof that a classical probabilistic model of quantum theory is possible; since Bohm had prior claim to this (with a different model), it was probably politeness that led the authors to concede this result to Bohm. In Chap. (9) I have argued that, nevertheless, Bohm's model did not achieve this goal.

As to their own work, the original papers of Wiener and Siegel contain an error in the calculation claiming to show that their procedure gives the same results as quantum theory. This statement is corrected in the book, in an appendix by Warnock, leaving some doubt whether this algorithm works in the sense claimed. Another criticism of the enterprise was made by Schwartz [66]. The sample space constructed by Wiener is the "differential space" attached to the points of the Hilbert space of the quantum model being discussed. The hidden variables are not, unfortunately, random variables on the sample space; for each ω, called α in [170], is constructed a set of observables (some of them hidden in quantum theory) whose values depend on the state being considered.

This is a very contextual model, in which the probability theory changes its observables, not with each maximal abelian set, but with each state. In a few paragraphs, with no mathematics, it is claimed that this can be patched up, but it is not very convincing. One of the troubles with patching it up is revealed in the next section on von Neumann's theorem on hidden variables, showing that quantum theory has no dispersion-free states. In the Wiener model, a point measure at any α is dispersion free, and it is suggested that these states might one day be found. The authors say that this shows that the Wiener-Siegel model does not satisfy the assumption of von Neumann, often criticised, that the mean of a sum of non-commuting observables is the sum of the means. We call this the linear axiom. The possible values of an observable X in the Wiener model are the eigenvalues of the quantum operator corresponding to X. The counterexample to von Neumann's linear axiom arises for the harmonic oscillator, in which $X = H = (p^2 + q^2 - 1)/2$. At a point α, one of the values of X occurs with probability 1. We can find a sample α where p^2 and q^2 take any positive values, but H takes only integer values. Thus, the linear axiom would fail for these states. Now, if the model can be fixed up to be a proper probability theory, such that the values of the random variables depend only on ω and not on the state, as claimed informally on p. 147, then von Neumann's linear axiom would hold (the state would be a linear functional) and these dispersion-free states could not exist. This seems to defeat the triumph of the model, the prediction of dispersion-free states. But all this tinkering with the model became a lost cause after Bell's theorem, in the version given in Chap. (8).

11.2 Nelson

Let us now turn to the various books by Edward Nelson. Nelson adopted stochastic mechanics as a dynamical theory of a particle governed by a classical stochastic process, which gives the same answers as quantum mechanics for the time-evolution of the probability distribution of the particle's position. Nelson discusses the extension of the theory to encompass two or more particles. He later abandonned it as not correct physics [125]. This has left a stranded group of enthusiasts, who continue to study the theory. The idea of studying two particles has been taken up in [121]. It seems to me that once we have two logically independent degrees of freedom, both with an underlying classical stochastic dynamics, then it is easy to find a statistical prediction of the theory that is not the same as that of quantum theory. We just need to set up four pairs of pairwise commuting observables, some of which are complementary as in Bell's examples. This would give us a Bell inequality for the Nelson theory, or ANY other description by a classical stochastic process, but not for the quantum theory. Therefore, there will be observable statistics in the classical theory that differ from those predicted by quantum theory. This convinces me that "it is not correct physics". Chris Weed has remarked that the recent preprint "Quantum Theory from Quantum Gravity", by F. Markopoulou and L. Smolin [116] seems to suffer from the same problem. The authors claim in the title and abstract that starting from a classical deterministic spin dynamics, coupled to a heat bath, one can perhaps get quantum mechanics. They try to avoid Bell's theorem by saying "The nonlocal hidden variables required to satisfy the conditions of Bell's theorem are the links ... in the graph". We should point out that the proof we gave of Bell's theorem makes no assumption about the hidden variables; the observables themselves are not assumed to have local dynamics; thus, there is no need for Bell's locality assumption; this was that the observables of one system do not depend on the hidden variables localised at the other; we just assume that the observables are random variables on a common sample space. That is, that their definition is non-contextual. The authors seem to be trying to construct a nonlocal quantum dynamics. There is no need to; it does not help, and anyway, relativistic quantum mechanics is, or should be, local. The authors final theory is not full quantum mechanics. It is a Nelson theory in which there is a wave function satisfying a version of Schrödinger's equation. The mod-squared wave function is the probability density that the particle is present at the time in question. However, as in Chap. (9) the velocity is not an observable incompatible with the position, but is the average of Nelson's forward and backward velocities. More, the positions of the particles at different times are random variables on the same sample space, to wit, the sample space of the driving noise of the stochastic dynamics. They have good, real correlations, unlike in quantum mechanics, in which the correlation between position at different times is complex. Since the dimension of the Hilbert space is at least four, there are pairs of compatible observables of Bell type which do not give the same results as quantum mechanics. The authors are not saved by any nonlocality of the dynamics.

Nelson's theory [123] is a substantial construction. It is more like the extension to Bohm's theory given by Bohm and Vigier in (1966) [27] than Bohm's original version [24]. Nelson [123] can be considered to be the rigorous version of Fényes. Let M be the configuration space of a system. A path is a function $\xi : \mathbf{R} \to M$. The velocity vector is then the function $\dot{\xi}(t)$. On page 60, Nelson starts with the principle that the dynamics is determined by the action I such that

$$I = \int_{t_0}^{t_1} L(\xi, \dot{\xi}, t)dt = \int_{t_0}^{t_1} L(\xi, v, t)dt \tag{11.1}$$

be stationary. This is Hamilton's principle; ξ is a critical path of I, under variations with the same initial and final positions. This leads to the *Euler-Lagrange* equations

$$\frac{\partial L}{\partial q_i} - \frac{d}{dt}\frac{\partial L}{\partial v^i} = 0. \tag{11.2}$$

Nelson's theory concerns a finite number of particles, whose mass is given in general coordinates by the mass tensor m_{ij}. Then we may define its kinetic energy by $T = \frac{1}{2}m_{ij}v^i v^j$, and the potential energy V by

$$L = T - V.$$

Nelson then notes that in normal coordinates q^i, $\partial T/\partial q^i = 0$. Then in normal coordinates, the Euler-Lagrange equation gives

$$-\frac{\partial V}{\partial q^i} - \frac{d}{dt}\left(m_{ij}v^j - \frac{\partial V}{\partial v^i}\right) = 0 \qquad \text{(normal coordinates)}.$$

But by Newton's law, $\frac{d}{dt}\left(m_{ij}v^j\right) = F_i$ in normal coordinates, so if F_i is the force,

$$F_i = -\frac{\partial V}{\partial q^i} + \frac{d}{dt}\frac{\partial V}{\partial v^i}$$

holds in normal coordinates. At this point, Nelson requires that the force should be a dynamical variable: if the point $\xi \in TM$ is known, and the time is known, then F_i must be known. Thus, $\frac{d}{dt}\frac{\partial V}{\partial v^i}$ must be independent of v. He assumes that this implies that $\frac{\partial V}{\partial v^i}$ in also independent of v, and gets the general form

$$V(x, v, t) = \phi(x, t) - A_j(x, t)v^j,$$

which, being a tensor equation, must hold in all coordinates. He thereby justifies the form for the force,

$$F_i = \left(-\frac{\partial \phi}{\partial q^i}\right) + \left(\frac{\partial A_j}{\partial q^i} - \frac{\partial A_i}{\partial q^j}\right)v^j.$$

We get, then, the equations of motion for a particle in a magnetic potential A_i as well as an electric potential ϕ. The idea of Nelson is that $\xi(t)$ should undergo a stochastic process such that the average of I should be stationary: $\delta \mathbf{E}[\int L\,dt] = 0$. He replaces the Hamilton principle (11.1) with its stochastic version, and shows that the mean energy is conserved in time, and the variation of the position and momentum of any particle obeys the Heisenberg uncertainty relation. Nelson finds a solution to this problem; he assumes that any measurement of position and momentum obeys

$$\mathcal{V}x\,\mathcal{V}v \geq \hbar^2/4, \qquad (11.3)$$

where $\mathcal{V}Y$ is the variance of the random variable Y. He says that a measurement of ξ will have an error δx, and that the measurement causes the particle to acquire an additional error in momentum of δp. He assumes that δx and δp are independent, with mean zero and that the position $\xi(t + dt)$ at a later time is disturbed to $\xi(t + dt) + (\delta p\,dt)/m + o(dt)$. This later position can be observed with great accuracy, since we are not interested in any disturbance of the particle after time dt. Then we have observed $\delta \xi(t) := \xi(t + dt) - \xi(t)$ to within an expected variance,

$$\mathcal{V}(\delta \xi(t)) = \mathcal{V}x + m^{-2}\mathcal{V}p\,(dt)^2,$$

since the cross-terms are zero. The minimal value of this error, subject to Eq. (11.3) is $\hbar \, dt/m$. He thus arrives at this value for $\mathbf{E}_t[(d\xi(t))^2] = \hbar \, dt/m$ at time t.

To study the case of random motion, one introduces a regular probability measure μ on the space of paths $t \mapsto \xi(t) \in M, t \geq 0$. For example, μ might be the Wiener measure. Let the *past*, $\mathcal{P}_t$, the *future* $\mathcal{F}_t$, and the *present* $\mathcal{N}_t$, be the σ−rings generated by the $\xi(s)$, with $s \leq t$, $s \geq t$ and $s = t$, respectively. Then one can define the conditional expectation at time t to be $\mathbf{E}_t = \mathbf{E}[\cdot | \mathcal{N}_t]$. See Chap. (3) for the construction of the conditional probability. Nelson defines the *forward differential* $df := f(t + dt) - f(t)$ for any function f, and any $dt > 0$. He also takes $\mathcal{B}_t$ to be the set of all uniformly bounded real stochastic processes $\eta(t)$, $\eta \in [0, \epsilon]$, $\mathcal{N}_\infty$−measurable, such that $\|\mathbf{E}_t \eta(dt)\|_\infty = O(dt)$, and $\|\mathbf{E}_t \eta(dt)^2\|_\infty = O(dt)$. Two such processes are taken to be equivalent if they differ by ζ with both these expectations $o(t)$ instead of $O(t)$. The set of equivalence classes is the stochastic tangent space at t of the set of processes ξ.

Nelson says that ξ is a *smooth diffusion* if in any neighbourhood U of a point, with local coordinates q^i, there exist smooth functions β^i and σ^{ij}, with σ of strictly positive type on U, such that

$$\mathbf{E}_t[d\xi^i(t)] = \beta^i(\xi(t), t)dt \tag{11.4}$$

$$d\xi^i(t)d\xi^j(t) = \sigma^{ij}(\xi(t))dt \tag{11.5}$$

and the same holds for the time-reversed process $\check{\xi} := \xi(-t)$. Nelson makes the assumption that the paths relevant to quantum stochasticity is a smooth diffusion. He notes that under change of coordinates, ξ^i is not a vector; this results from the Ito lemma, which is Eq. (5.4) in [123]:

$$df(\xi(t), t) = \frac{\partial f}{\partial q^i}(\xi(t), t))d\xi^i + \frac{1}{2}\frac{\partial^2 f}{\partial q^i \partial q^j}(\xi(t), t)\sigma^{ij}(\xi(t))dt$$

$$+ \frac{\partial}{\partial t}f(\xi(t), t)dt. \tag{11.6}$$

He notes in Eq. (5.5) that σ^{ij} is, however, a tensor of rank 2, when we change the coordinates.

Nelson defined the backward derivative of a process to be

$$d_* f(t) := f(t) - f(t - dt),$$

and the forward and backward stochastic derivatives to be

$$DF(t) = \lim_{dt \to 0_+} \mathbf{E}_t \frac{F(t + dt) - F(t)}{dt} \tag{11.7}$$

$$D_*(t)F(t) = \lim_{dt \to 0_+} \mathbf{E}_t \frac{F(t) - F(t - dt)}{dt}. \tag{11.8}$$

Nelson then presents his results, modified by the work of Guerra and Morato, and of Yasue. Let us give Yasue's version. Let L be the Lagrangian, of the form deduced above. Let

$$J(\xi) := \mathbf{E} \int_{t_0}^{t_1} \left[\frac{1}{2} L(\xi(t), D\xi(t), t) + \frac{1}{2} L(\xi(t), D_*\xi(t), t) \right] dt. \qquad (11.9)$$

Nelson asks for the ξ that are *critical for L*. This means that ξ has finite energy, and that for all time intervals $[t_0, t_1]$ and time-dependent smooth vector fields X with compact support in $M \times (t_0, t_1)$, we have

$$J(\xi + X) - J(\xi) = o(X). \qquad (11.10)$$

He then proves Theorem (14.2), which states that if ϕ and A_i are smooth with compact support in (t_0, t_1), then a smooth diffusion ξ of finite energy is critical for L if and only if the stochastic Euler-Lagrange equations hold:

$$\frac{\partial L}{\partial q^i}(\xi, D\xi, t) + \frac{\partial L}{\partial q^i}(\xi, D_*\xi, t) - D_* \frac{\partial L}{\partial p^i}(\xi, D\xi, t) - D \frac{\partial L}{\partial p^i}(\xi, D_*\xi, t) = 0. \qquad (11.11)$$

The upshot of this result is that $m_{ij}a^j = F_i$ holds, where a^i is the stochastic acceleration of the field ξ, and F^i is the force. Note that we do not get that ξ obeys the Markov property.

Nelson goes on to show that the solution $\xi(t)$ allows one to solve the Schrödinger equation, to get $\psi = e^{R+iS}$, in which $\rho(\mathbf{x}, t) = |\psi(\mathbf{x}, t)|^2$ is the correct particle density for all time. Thus the theory gives the same results as the corresponding Schrödinger quantum theory, provided that you do not attempt to measure the position of the particle at two different times. He says on page 118 that to do this would need measuring instruments that are not at equilibrium with the system. This possibility is not ruled out by his book. We see from the usual form of the Lagrangian, for example, $L(x) = \frac{1}{2}\dot{\xi}^i\dot{\xi}_i + \phi - A_j p_j$, that its integral involves $\mathbf{E}[\int \dot{\xi}^i \dot{\xi}_i dt]$. Since we have, in stochastic analysis, that $d\xi^i d\xi^j = \sigma^{ij} dt$, we see that $\dot{\xi}^i\dot{\xi}^j = \sigma^{ij}/dt$; the integrand, together with dt, then comes to σ^{ij}, with the dt cancelled out: there is no way the integral can make sense. Nelson shows, however, that in the variation of the integral, the singular term is the same in the two terms $J(\xi)$ and $J(\xi + X)$, and so cancels out. This is how Nelson gives a meaning to the theorem (14.2).

We now know (see Chap. (9)) that there are some predictions of Nelson's theory that are not in conformity with quantum theory. This was not enough to make Nelson give it up. But finally, on page 127, he abandons the Markov principle, on the grounds that it leads to a nonlocal theory. Unlike Bell, who advocated the rejection of special relativity, Nelson wanted to keep it. To do this, we need to abandon particles, and describe the system by fields. Thus, Nelson gives up.

12

Other Lost Causes

12.1 Bondi's Solution to the Twin Paradox

Two twins embark on the following idea. One twin, A, remains on earth and gets older using a clock measuring time t. The other, B, flies off on a voyage of speed v relative to A. According to special relativity, his clock shows time-dilation, being related to t by

$$t' = t \left(1 - \frac{v^2}{c^2} \right)^{1/2} .$$

At a certain distance away, B reverses his velocity and travels back to A at the speed v. On the home trip, his time is still related by the same formula to A's time, so when he gets home he finds that his twin has aged relative to himself. However, in general relativity both observers are equivalent in related coordinates, and B sees A go away, come to rest and return to him at speed v; thus A would age faster than B from the point of view of B, and we have a contradiction.

Bondi solved this paradox by asserting that the twin who had suffered the change in speed will be the younger of the two. He gave the following neat argument. Suppose that B moves from earth to Syrius at a constant speed v, holding his clock showing time t'. At Syrius, B does not stop, but passes an observer C moving back to earth at a speed v. C is holding a clock showing exactly the same time as B's clock shows at the instant they meet. Namely, relative to A, C's clock shows t'. This time is then less than t, so when C meets A, he is younger. Only special relativity is needed, and the symmetry between the twins is broken: A meets C later than B met him.

12.2 Quantum Logic

This subject was invented by von Neumann, Jordan and Wigner. They devised
an algebraic set of axioms for propositional logic, different from the Boolean
algebra of classical logic. The Boolean algebra of a set A is the collection of
its subsets, including the whole and the empty set; the collection of subsets
is called the *power set* of the given set A. The power set is furnished with
a *sum*, the union of two subsets minus their intersection, and a *product*, the
intersection of two subsets. This structure forms a ring in modern parlance;
it obeys the distributive property. It is an algebra only in the formal sense
that any ring is an algebra over the trivial field containing 0 and 1. In contrast, the quantum logic of von Neumann and friends is a non-distributive
lattice. There is really only one result in the subject: Piron's theorem, which
says that (subject to the covering property, and some regularity assumptions
like measurability) any quantum logic is isomorphic to the lattice of subspaces of a Hilbert space over some field. This leads to the idea of replacing
von Neumann's complex Hilbert space with real quantum mechanics, when
the field is **R**, or quaternion quantum mechanics, when the field is the non-
commutative field of quaternions. One might even try p-adic fields. So far,
nothing of physical importance has arisen from these attempts, except possibly the quaternionic case. The same can be said about attempts to generalise
the logic even more, in an attempt to avoid Piron's theorem.

Omnès [131] says that this was not a very good idea. Instead, he shows how
to argue using classical logic, within quantum probability. This is the version
of Griffiths's logical histories preferred by Omnès. Griffiths starts with the
facts: suppose that non-commuting projectors P_i, $i = 1, 2$ are measured at
times t_i, $t_1 < t_2$. If the initial state is ρ, then after both measurements we get
from von Neumann's formula the state

$$P_2 P_1 \rho P_1 P_2 + (I - P_2) P_1 \rho P_1 (I - P_2) + P_2 (I - P_1) \rho (I - P_1) P_2$$
$$+ (I - P_2)(I - P_1) \rho (I - P_1)(I - P_2).$$

If the observed results are $P_1 = 1$ and $P_2 = 1$, then the probability of this is

$$\mathrm{Tr}\,[P_2 P_1 \rho P_1 P_2] = \mathrm{Tr}\,[P_1 \rho P_1 P_2^2] = \mathrm{Tr}\,[P_1 \rho P_1 P_2].$$

These are facts. However, we might ask what would have happened if instead
of P_2, the second experiment had been to measure P_3? Clearly, it is consistent
to say that the probability would have been got with P_3 replacing P_2. This
is found to be true experimentally; this arises because the last expression is
linear in P_2.

The measurement of P_3 is counterfactual: it did not occur. But we can ask,
what is the probability of getting $P_3 = 1$ *if* we had measured it second, instead of measuring P_2? We get a consistent reply in this case. For more general
situations, we cannot ask what would have happened in a counterfactual situation. However, Griffiths (and Omnès) tell us when a counterfactual question

can be asked; this is true when ρ and the projectors satisfy certain properties, discussed in Eq. (6.6) and called the Griffiths consistency conditions.

12.3 Trivalent Logic

Quantum logic is criticised for not being a logic in the book, [68], by Rachel Wallace Garden. She points out that the propositions of a (classical) logic should admit valuations, which is an assignment of truth or falsehood to each proposition. The lattice approach leads to a structure for which valuations might not be meaningful, and for which the distributive property fails. The book describes R. Garden's approach; she allows a three-valued valuation, true, false, undecided. But she maintains the distributive property, and so remains close to a classical theory. Quantum theory fits into the general scheme; for her, a "state" is, in physicists' language, a pure state, the simultaneous eigenstate of a maximal abelian subalgebra of the observables. This abelian algebra defines the measurement, M, being done. Such a state defines a probability space, giving the same distributions as predicted by quantum theory. The sample space of this is contextual, in that a different choice of M gives a different sample space. The construction of the sample space is via the Stone space of the Boolean 'algebra' defined by M. In finite dimensions, this is isomorphic to the space constructed using the Gelfand isomorphism; see Chap. (3). Her conclusion, that the theory leads to a Kolmogorovian probability theory, is bizarre; the same observable is realised by various different random variables; even the sample space depends on which (compatible) further observables are being measured. It is clearly a frequentist theory, very prekolmogorovian: apart from the introduction of various sample spaces for the same observable in different contexts, there is no general requirement taking the place of Kolmogorov's consistency condition. For the examples of classical and quantum probability, this requirement is satisfied in her set-up. However, it is difficult to see how any other case of the general formalism could be constructed obeying the consistency condition. For example, it is not required that a given observable have the same distribution when it is regarded as a random variable on different spaces. The author is perplexed by the theorem of Kochen and Specker, which has similar consequences for her as Bell's theorem would have had, if she had known about it. She speculates that perhaps its proof is not correct; this speculation is, of course, unfounded.

12.4 Jordan Algebras

Jordan introduced a real algebra that was not associative, but which satisfied an equation of the fourth degree, as a model that includes the hermitian matrices as a special case. The motivation was to take the product of observables A and B as

$$A \circ B := \frac{1}{2}(AB + BA). \tag{12.1}$$

If A, B are matrices, then this product maps the self-adjoint subset to itself. Jordan noticed that it also satisfies

$$((A \circ A) \circ B) \circ A = (A \circ A) \circ (B \circ A) \tag{12.2}$$

for all $A, B \in \mathbf{C}^n$. Indeed, by using Eq. (12.1), both sides of Eq. (12.2) are easily shown to be equal to $\frac{1}{4}(A^2BA + BA^3 + A^3B + ABA^2)$. Jordan then took the quantum algebra to be a real commutative algebra obeying Eq. (12.2). The product given in Eq. (12.1) is obviously commutative; but it fails the associative property, in that there exist matrices A, B and C such that $(A \circ B) \circ C \neq A \circ (B \circ C)$. Thus, Jordan omitted associativity from his axioms. One can put a norm on a Jordan algebra, and complete to get a Jordan$-C^*-$algebra, and try to do quantum mechanics. This generalises the C^*-algebra theory of Segal. Indeed, Paul Cohn showed that a Jordan algebra with a certain property was the hermitian part of a C^*-algebra; the Jordan product $\circ$ was related to the associative product of the C^*-algebra by Eq. (12.1). Such a Jordan algebra is said to be *special*. The trouble with non-special Jordan algebras is that there is an example containing the eight by eight matrices with octonain entries; this has no representation by complex matrices (which would be associative, which the Jordan algebra is not). As a consequence of this example, there is no way, from two general Jordan algebras, to form their tensor product; the tensor product of their underlying vector spaces is doable; the problem is, how to define a Jordan product on it. This is needed in any theory used to describe local systems. We conjecture that if $\mathcal{A}$ is a Jordan algebra, with the property that a tensor product with the Jordan algebra of the complex matrix algebra $\mathcal{M}^2$ can be defined as a viable Jordan algebra, then $\mathcal{A}$ is a special Jordan algebra; that is, it is isomorphic to the hermitian elements of a C^*-algebra. Because of this problem, we must discard Jordan's nice theory.

Experience shows that it is more productive to use classical logic, but to change the probability theory from classical to quantum. Then we may replace the algebra of all bounded operators on a given Hilbert space by a more general C*-algebra, and still be able to do physics. See [81], Haag's book, **Local Quantum Physics**. Very little physics has resulted from quantum logic, trivalent logic, or Jordan algebras.

12.5 Non-self-adjoint Observables

It has been suggested that the concept of observable in quantum mechanics might be extended to include, not only self-adjoint operators, but also any operator similar to a real diagonal operator [21]. That is, A is observable if there exists a similarity transformation S, with densely defined inverse, such that $S^{-1}AS$ is real and diagonal. The idea comes from the easy fact that such operators, which we shall call diagons, have real spectrum. If we postulate, with von Neumann, that any measured value of an observable is an eigenvalue, or more generally, lies in the spectrum, then allowing such operators will not lead to any complex measured values. Of course, self-adjoint operators are diagons, so the proposal is an extension of the set of observables in quantum mechanics. It has also been proposed that any operator with real spectrum might be included as an observable. Is there anything theoretically wrong with it, and if not, why was it not discovered by the people working in quantum logic and Jordan algebras?

The first thing tried was to expand the allowed dynamics so as to allow a non-self-adjoint diagon as the energy operator, and to ask for the consequences. Unless we abandon the relation between the energy and the generator of time-evolution, which is the most successful idea in classical as well as quantum theory, we are then led to a non-unitary evolution. This was the point of the foundational work of Misra and Prigogine: in a dissipative system, we need a contraction semigroup, while the observed values of the energy H must be real. Indeed, a semigroup occurs naturally as the linear approximation to my own nonlinear heat equations, paper [158]. Where this work differs from the scheme of Misra and Prigogine is that, as in any good theory of non-equilibrium thermodynamics, what is dissipated in [158] is not energy, or total probability, but information. In [158] total probability is conserved in time, and energy is simply transferred from one sort to another, such as from potential energy to heat. The mean total energy is conserved (the first law of thermodynamics). A dynamics in which the total energy leaks away to zero would obviously violate this law. For this reason, choosing a diagon as Hamiltonian does not seem to be a good idea. This relates to another question, that of transition probabilities, which is given by the mod square of the scalar product of the two states in question. Since the eigenvectors of a diagon H in general are not orthogonal, there will be a non-zero transition probability that the observed energy of an eigenstate will be higher or lower than that of the initial state. This cannot be a good fundamental theory; it looks more like a system with random noise.

Carl Bender and coworkers were led to study diagons from analytic perturbation theory. They have tried to make it into a valid generalisation of quantum mechanics. See [21]. This and possible lines of further work are also described in [34].

Here, the authors consider the usual Schrödinger operators p and q acting on wave-functions of q, and build Hamiltonians that are not hermitian as operators on $L^2(\mathbf{R})$, but which are invariant under an antiunitary operator identified as PT (parity times time-reversal). They argue that such a Hamiltonian not only has real spectrum (which is proved), but is also very likely to have a complete set of eigenfunctions (which has been verified to great accuracy by numerical studies). They do require that the boundary conditions are carefully chosen in the eigenvalue problem. It is not clear under what conditions the eigenfunctions are square-integrable. The eigenfunctions split into pairs, related by the PT transformation and they define a natural indefinite scalar product. This is made positive definite by changing the sign of the 'norm' on half the basis. This operation is likened to the charge operator C and Dirac's trick with the positron states: the scalar product uses an analogue of the PCT conjugation in place of complex conjugation. Thus they end up with a Hermitian operator H as Hamiltonian relative to this new scalar product. All the eigenfunctions have finite norm in the new setting. In the models discussed, various anharmonic oscillators with complex potentials, neither p nor q commutes with H. It would appear that they are not observable, being non-hermitian in the new scalar product. An interesting problem (but not one for a Ph. D. student) is whether a viable two-particle theory can be constructed. We need a Hermitian operator to act as the position of the centre-of-mass of the two particles; this will generate the symmetry group of Galilean boosts. We need another, the total momentum operator, to generate translations. Their commutation relations with H are also prescribed by the rules of the Lie algebra. Then, they must give one of the known reducible projective representations of the Galilean group, perhaps leading to a theory in which the two particles do not possess individual observable positions and momenta. My guess is that it cannot be done. As it is, the work leads to a new idea for interacting particles, which however does not lead to finite results for interacting fields. The problem is, that the integrals occurring diverge even worse than in Feynman's theory! This is shown below, in Eq. (12.6).

One of von Neumann's axioms, used to exclude hidden variables in quantum mechanics, is that the sum of two observables should be an observable. For example, if one's hidden variables are to be random variables, and all random variables are observable, then this axiom is true. von Neumann allowed a more general scheme, in which the quantum self-adjoint observable could be random; his axiom is also true in this case. Nevertheless, this axiom has been criticised, almost derided, by Bohmians, as unwarranted: if two observables are not simultaneously measurable, there is no operational (that is, experimental) definition of their sum. Their means can be summed, but there might be no observable of which this is the mean. Diagons provide a concrete model in which von Neumann's axiom is violated: if A and B are diagons, then in general $A + B$ is not a diagon. This reveals the basic problem: the expectation values of a diagon in all (pure) states are real only if it is self-adjoint. Thus we would be motivated to assume that a diagon is observable only in states in which its mean is real, and this is usually only its eigenstates. Then the question would arise, suppose that we have an apparatus for measuring the observable, A, and we feed it a state for which the mean is complex: what would our apparatus do?

If we have only one observable in a theory, and it is a diagon, then we can make it self-adjoint by changing the metric in the Hilbert space, so that its eigenvectors are mutually orthogonal. This was remarked by Bender and friends for their class of models. If we have more than one observable, however, we can play this trick only among those observables that mutually commute. The scalar product between two Hilbert space vectors would then depend on what set of mutually compatible observables we are thinking of measuring, and so the transition probability would have no invariant meaning.

In a quantum theory with a diagon as energy, we lose Wigner's theory of symmetry; the observables cease to be a vector space, which means that we are not allowed to perturb a diagon with any old diagon; and the energy will be observable only in some states. Not much is left.

See also href=http://arxiv.org/abs/quant-ph/0310164, which is a Critique of PT-Symmetric Quantum Mechanics, by Ali Mostafazadeh.

In a later paper [22], Bender, Brody and Jones take the view that the scalar product defined for the anharmonic oscillator might have a generalisation to quantum field theory. They choose to renounce the usual scalar product, and use the one such that H is hermitian relative to it.

Using an algebraic method, this has been carried out in some detail. They work out the case

$$H = \frac{1}{2}p^2 + \frac{1}{2}\mu^2 x^2 + i\epsilon x^3. \tag{12.3}$$

The eigenvalue E_n and eigenstates $\phi_n(x)$ satisfy the Schrödinger equation

$$-\frac{1}{2}\phi_n''(x) + \frac{1}{2}\mu^2 x^2 \phi_n(x) + i\epsilon x^3 \phi_n(x) = E_n \phi_n(x). \tag{12.4}$$

The appropriate boundary condition is $\lim_{|x|\to\infty} \phi_n(x) = 0$. It was known that this operator has positive spectrum [39]. In [22] they use that the positivity of H is associated with a spacetime reflection symmetry $\mathcal{PT}$; this means that every eigenstate of H is an eigenstate of $\mathcal{PT}$. More, the authors show that there is a scalar product on the span of the space of eigenfunctions of H, such that H is symmetric on the Hilbert space constructed out of this. The clue is to start with the expression

$$C(x,y) = \sum_n \phi_n(x)\phi_n(y).$$

They then show that the bilinear form

$$\langle \psi, \chi \rangle := \int dx\, \psi^\Theta(x)\chi(x)$$

is positive definite on the space, and that H is self-adjoint. Here,

$$\psi^\Theta(x) := \int dy\, C(x,y)\psi^*(-y).$$

This construction clearly depends on H.

The trouble is, one needs to find $\phi_n(x)$ to proceed; in [22], the authors suggest an approximation method which avoids this problem. They write

$$C(x,y) := \exp\{\epsilon Q_1 + \epsilon^3 Q_3 + \cdots\}\delta(x+y) = e^{Q(x,p)}\mathcal{P}.$$

Here, Q_{2n+1} are differential operators depending on x and $p = -i\frac{d}{dx}$. In this form, only odd powers of ϵ enter, and their coefficients are real. One shows that $Q(x,p)$ is even in x and odd in p. The Hamiltonian (12.3) can be written $H = H_0 + \epsilon H_1$, where H_0 is the harmonic Hamiltonian that commutes with $\mathcal{P}$, and H_1 is cubic in x, and anticommutes with $\mathcal{P}$. From these properties, they get

$$2\epsilon e^{Q(x,p)}H_1 = [e^{Q(x,p)}, H] = e^{Q(x,p)}H - He^{Q(x,p)}.$$

We expand Q in powers of ϵ, and collect the coefficients of like powers of ϵ; then we get a family of equations for the coefficients:

$$[H_0, Q_1] = -2H_1$$
$$[H_0, Q_3] = -\frac{1}{6}[Q_1, [Q_1, H_1]]$$
$$\cdots$$

It turns out that we can solve these in sequence if we assume that Q_{2n+1} is a polynomial in (p, q) of maximum degree $2n + 3$. They then give us [22] the first four terms; for example,

$$Q_1 = -\frac{4}{3}\mu^{-4}p^3 - 2\mu^{-2}xpx.$$

The authors then try to do the same for the quantum field theory with Hamiltonian

$$H = \int dx \left[\frac{1}{2}\pi_x^2 + \frac{1}{2}(\nabla\varphi_x)^2 + \frac{1}{2}\mu^2\varphi_x^2 + i\epsilon\varphi_x^3\right].$$

They claim that it goes through; however, I notice that the equation (17) of [22] for the coefficient M in the expression for the first approximation, Q_1 diverges; indeed,

$$Q_1 = \int\int\int dx\,dy\,dz\,(M\pi_x\pi_y\pi_z + N\varphi_y\varphi_x\varphi_z), \tag{12.5}$$

where M is given by

$$M = -\frac{4}{(2\pi)^3}\int\int dp\,dq\frac{e^{i(x-y)\cdot p+i(x-z)\cdot q}}{4[p^2q^2 - (p\cdot q)^2 + O(p,q)^2]}. \tag{12.6}$$

This equation (12.6) diverges to infinity, whereas it should have been smooth for Eq. (12.5) to make sense; indeed, the numerator contains $dp\,dq = |p|^2|q|^2 dp\,dq$, of fourth degree in momentum, which cancels with the same expression in the denominator. One is left with $\int dp\,dq$, which diverges quadratically. So it is back to the drawing board.

12.6 Critique of Geometro-Stochastic Theory

In his book on geometro-stochastic quantum mechanics, Prugovecki argues forcefully for a new approach to fundamental physics. He quotes extensively from the masters, and also from lesser people, and has a wonderful list of references. Along the way, he offers criticism of present theories and existing trends. To make the case for his new theory, abbreviated *GS* in the following, he is obliged to say something about the other theories used by particle physicists: his theory is not is accordance with what Haag calls "Einstein causality", and Einstein is one of Prugovecki's heroes. I shall not explain, even if I could, what *GS* theory is; for that, buy his books! Rather, I shall try to answer his remarks about local quantum field theory and Haag's algebraic programme.

Prugovecki argues that local commutativity "has no physically truly meaningful relationship to the question of Einstein causality any more than ... in classical relativistic theory. Indeed, in classical special relativistic theory, the commutativity of all observables is trivially satisfied". Here, Prugovecki has put his finger on the difference between the two concepts that Haag [81] calls "Einstein causality" and "primitive causality". I believe that Prugovecki has confused the two. According to Haag, local commutativity is needed so that a measurement of an observable localised in a space-time region $\mathcal{O}$, and the consequent reduction of the state, causes no instantaneous influence in a region $\mathcal{O}'$ space-like to $\mathcal{O}$. In the famous paper, R. Haag and D. Kastler [82] outline a proposal as to how a local measurement can be achieved mathematically. I now present a modern version of this.

We measure A in the region $\mathcal{O}$ by applying a completely positive stochastic map, denoted $T(A)$, to the algebra of all observables, with the properties that on the local algebra in $\mathcal{O}$, the dual $T^*(A)$ of $T(A)$ agrees with the conventional result of a measurement on the vector-states; that is, it gives the reduction of the wave-packet, into the mixed state consisting of eigenvectors of A, weighted with their transition probability from the initial state. On any observable B located space-like to the space-time region involved in measuring A, $T(A)$ has no effect: $T(A)B = B$. It follows that the measurement of A has no effect on the states observed in the region where B is localised. Later work showed that local commutativity, namely the condition $[A, B] = 0$, is a necessary ingredient for the general existence of such $T(A)$. This was explained in Chap. (6). Notice that if the observer in $\mathcal{O}$ not only triggered the measuring apparatus for A, but also read the result of the measurement, then the reduced state on the algebra in $\mathcal{O}$ would become conditioned, and only the eigenprojection with the observed eigenvalue would remain in the state at $\mathcal{O}$. This knowledge is not available to the observer at B, so the state on the algebra at $\mathcal{O}'$ is still unaffected by the measurement. If $\mathcal{O}$ communicated the result to $\mathcal{O}'$ by a signal, then this information would change the state at $\mathcal{O}'$, and it would collapse; this can't happen if they are space-like separated. To keep track of which state is assigned by whom, we must introduce *information sets* and treat the problem as in game theory; the algebras generated by the indicator functions of the information sets could be called "information algebras". This can be generalised to non-abelian algebras in the theory of quantum games. This is suggested in my book [157]; a brief account is to be found in Chap. (3) and Chap. (4).

Why did Haag [81] call this "Einstein causality"? I surmise that the term is short for "Einstein-Podolski-Rosen causality". In their *EPR* paper, Einstein, Podolski and Rosen suggest a thought experiment, which, they argued, showed that quantum mechanics is not "complete". Part of their argument assumes that any measurement of one observable, such as A above, will also reduce the quantum state of an observer looking at B. They were forced to this conclusion, because mixed states (that is, statistical mixtures represented by density matrices) were excluded, following Einstein's dictum, "God does not play dice". Thus arose a misconception, wherein the "influence" of a measurement was alleged to travel faster than light. Thus, in "The Emperor's New Mind", R. Penrose almost asserts that the experimental set-up used in $\mathcal{O}$, namely, which observable is being measured, but not the result of the experiment, is instantaneously transmitted to $\mathcal{O}'$. This is not what quantum theory says. The simple non-relativistic model of two particles of spin 1/2, uses a four dimensional Hilbert space H, the tensor product of two copies of spin$-1/2$ spaces; the *EPR* state is an entangled pure state on the algebra of all observables, but is a fully mixed state (2×2 density matrix, equal to 1/2 of the unit matrix) when viewed by the observer of A, and also when viewed by the observer of B; and the spin observables of one particle commute with those of the other. This commutativity allows the direct construction of the map $T(A)$ for any spin observable A of the first particle, and having the above property of not changing the state of the second. The requirement that the effect of an observation in quantum mechanics does not travel faster than light can (arguably) be said to have originated in *EPR*, and so might deserve the name "Einstein causality".

The other concept, primitive causality, originated in the even earlier paper, that of R. Haag and B. Schroer [83]. It postulates that the observables localised in a region $\mathcal{O}$ generate the observables in the double cone causally defined by $\mathcal{O}$. This is the quantum version of the concept of "domain of dependence" which holds for classical fields obeying hyperbolic partial differential equations; it requires (in classical physics) that nonlinear terms be local, and that the order of the equation be finite. Then the characteristics of the equation define the "light cone" used to define the domain of dependence (and the range of influence). It is clear that Prugovecki would have preferred this idea, rather than local commutativity, to have been called "Einstein causality"; primitive causality does not involve any measurement; it is designed to imply that nothing can move outside the range of influence, and in particular, for the relativistic wave equations, nothing can move faster than light. Although Einstein can be said to have stimulated ideas related to this classical concept, he did not say anything about it in the quantum context, unlike his input into *EPR*.

Returning to *GS*, Prugovecki introduces a fundamental length (the Planck length), and abandons local commutativity. He says "Contemporary experimental high-energy technology ... is still far from being able to probe energies and distances of 'Planckian' orders of magnitude ... Hence the choice between conventional models and their *GS* counterparts is not one that could be made ... on the basis of experiment alone." This bold statement is a hostage to fortune; experimentalists are very clever people, and might be able to devise an experiment to detect the violation of "Einstein" causality, inherent in *GS*. One only needs to recall Salam's model, in which the proton was assumed to be unstable, with a life-time of 10^{31} years. Salam presented this model at the Rutherford Lab., on the occasion of Dr. Stafford's 25th Christmas conference. Salam said in his talk that this hypothesis had no experimental consequences: no-one was going to be able to watch a proton for that long. Nevertheless, at Kamiokande, people watched 10^{34} protons for 10 years, and none decayed; so the lifetime had to be revised upwards by four orders of magnitude, and it no longer served any purpose in Salam's model.

Closer to the question, is local commutativity exact, or is it violated at Planck lengths? is a very recent observation in astronomy. A supernova, which exploded maybe 4 milliard years ago, was observed; the photons of the various colours all arrived at the same time, within the time-span of the explosion. I would have expected this, since all photons travel at the same speed. However, the BBC report on the story said that "physicists" expected them to arrive at different times, because "they" believe that photons of different colour travel at different speeds. Who are these physicists? The story starts with the work of Amelino-Camilia [3] (available at the archive, gr-qc/0012051). This analysis suggested that Einstein's relation between the mass, energy and momentum of a particle (the *dispersion relation*) might be changed. The theory was called "double relativity", because not only is the speed c the same in all inertial frames, but also Planck's length is the same: it is the unique length that does not suffer from the Lorentz contraction. Various models were constructed, in one of which a massless particle might have a speed greater than c, depending ever so slightly on its colour; then c is the speed of photons in the limit of long wave-length. This was taken up by J. Magueijo, whose book, [115] was fortuitously published at the same time as both the BBC report, and an article in the New Scientist on double relativity. The author was confident that there would be no way to test the theory for several years, until a new facility in astronomy is finished; measuring the speed of light in the lab is not accurate enough. Nevertheless, the theory was refuted in quick time, the same week in which the book appeared, by the astronomical event mentioned.

Prugovecki makes a valid criticism of Haag's local quantum physics, which postulates that to any open set of space-time there is associated an algebra of observables; more, the operators assigned space-like separated regions are postulated to commute. Prugovecki points out that to probe small distances needs high energy, and that this cannot be made available by a small piece of apparatus lying in the region in question. It had been long understood that the operators assigned to an open set $\mathcal{O}$ of space-time were to represent the observables that could be measured by a piece of apparatus, switched on for a small interval of time, with the spatial extent of the apparatus lying in the space part of $\mathcal{O}$ during the time-interval needed for the measurement. This *raison d'être* for Haag's theory has been advocated by Borchers. However, looking at a typical CERN measuring device one sees that to probe very small distances needs enormous bits of hardware. Thus, Prugovecki argues, Haag needs classical and non-relativistic concepts to give experimental meaning to localisation in small regions. In particular, the apparatus we use certainly interferes with the environment outside the causal domain of influence generated by the local observable being looked at. He concludes that local commutativity of space-like observables might not be valid for distances of the size of the Planck length.

In answer, we must admit that we cannot be sure that an observable assigned to a small region $\mathcal{O}$ can be observed from inside $\mathcal{O}$. For example, who can say that we can measure the electric field, smeared with a test-function with support inside a region of space-time the size of a Planck volume, with a piece of apparatus smaller than the earth? We must provisionally weaken Haag's postulates. We could then understand that an observable A assigned to $\mathcal{O}$ can be observed by a large apparatus, but which still occupies a finite volume of space-time. Since the measurement cannot interfere with any other, B say, made at a large enough distance, we must then postulate that the commutator $[A, B]$ vanishes for all B located far enough away. In *GS*, such commutators, albeit very small, are not zero. Prugovecki claims that no experiment in the foreseeable future will be able to detect this violation of (Haag's) Einstein causality, or to show that it does not happen. Perhaps a clever astronomer will be able to show this to be too pessimistic.

12.7 Dirac's Programme of Quantisation

In the early days of quantum theory, Dirac proposed that the quantum analogue of the Poisson bracket of two classical observables should be the commutator of the corresponding quantum observables. This leads to the Dirac quantisation rule: to each classical observable A should be assigned an operator $q(A)$, such that the map $A \mapsto q(A)$ is a homomorphism of Lie algebras. This means that the classical observables must form an algebra with bracket the Poisson bracket, and the operators form a Lie algebra with the commutator as bracket.

It turns out that if one wants an irreducible representation of the Poisson algebra by operators, then it only works for some sub-algebras of the set of smooth functions of canonical variables, p, q. Of course, this includes the quadratic algebra in p and q, which is enough to describe the energy of the harmonic oscillator; this might have been the inspiration for Dirac's idea. Indeed, in the case of the quadratic algebra, the idea works for any number of degrees of freedom, including quantum fields (free ones). This leads directly to Irving Segal's real wave and complex wave realisations of the free field [147].

Attempts to extend the idea to polynomials of higher degree led to contradictions with the requirement that the map be a homomorphism; the quantum operators do not commute, so there is some ambiguity in defining a polynomial of p and q in the quantum case. Groenewald and van Hove showed that there is no ordering convention of the quantum operators that works for all polynomials. The most convincing result is that of Lionel Wollenberg: there is an algebraic property of the Poisson algebra that is invariant under Lie homomorphisms, but which is not shared by the quantum algebra of operators [173]. Nevertheless, there are **STILL** books on quantum mechanics being published which state that Dirac had solved the problem of finding a quantum theory associated with a given classical theory. Maybe this is what Bert Schroer is complaining about in his article [144], in which he (Schroer) denounces "quantisation": quantum theory has its own way of dreaming up models, and should not have to rely on classical models plus quantisation.

A more lenient approach was invented by J.-M. Souriau, in his brilliant book, [150]. He relaxed the requirement that the quantum algebra be irreducible, and was able to find a class of homomorphic mappings from the Poisson algebra to the algebra of linear operators on the space of smooth functions on phase space. Independently, in 1964 Irving Segal constructed a realisation of the quantum field algebra as (Gateau) differential operators on the linear space of smooth functionals on the Hilbert space of one-particle states. Segal did not require that the map be defined for all polynomial functionals (or tame functions), and his formula differs from that suggested by an extension of Souriau's formula to infinite dimensions. This was the subject of my paper, [154] written while I was Segal's research assistant at MIT. I found that Segal's map did not preserve the Lie structure, and that his claim that the action of the Lorentz group on the classical phase-space was mapped into the Lorentz action on the quantum operators was not true. Segal was kind enough to give me his hand-written notes of the calculation he had done, which convinced him of his claim. He had had extraordinary bad luck in this calculation. I remark on page 370 of [154] that Segal's quantisation differs from that of Souriau in eq (53) by the omission of the terms involving the action $S(t)$. In his check that the generators of the Lorentz group in his quantisation satisfy the correct Lie-algebra relations, Segal had arrived at an expression that should vanish if his theory were Lorentz invariant. This expression, which filled a page, arose exactly from the omitted term in (53), which came in twice. The second time it entered it was multipled by the sign of -Trace(G), where G is the Lorentz metric. This is +1 if we use (as Segal had done from the outset of his calculation) the time-like convention for the metric. Then we get a non-vanishing result, and Lorentz invariance fails. But Segal, seeing that all the 30 or so terms would cancel, term by term, if he took this trace to be positive, forgot that he had a time-like convention, and happily cancelled out all Lorentz-violating terms; this he considered such a remarkable chance that he was convinced that the theory was right.

Souriau theory is now called "prequantisation" rather than geometric quantisation, its early name. As I show in [154], it is not possible to find an invariant subspace of the space of functionals, on which the energy is bounded below, for a nonlinear quantum field theory. Kostant has been able to make prequantisation work when the observables form a solvable Lie algebra of type I, by the method of "holomorphic induction". This reduces to the Bargmann-Segal space of holomorphic functions on phase space for the harmonic oscillator. This was the subject of the other paper I wrote while at MIT, [155].

G. G. Emch has been able to identify the energy in some of these representations as the generator of the dynamics in a *KMS* representation; this allows non-positive operators to arise as generators of the time-evolution group. This, of course, violates the stability requirement, and cannot be allowed in an irreducible representation of the algebra. It is all right for the equilibrium states at positive temperature, as used by Emch.

12.8 *Diff M* as a Gauge Group

In quantum cosmology, it is often asserted that all observables are invariant under the diffeomorphism group of the space-time. This has led to a fantastic claim by Julian Barbour that one can remove time as a variable in cosmology. This is countered by Stuart Kauffman and Lee Smolin in an article [99], where the authors mention that one possible weakness in Barbour's argument is that he is dealing with spaces that are not algorithmically classifiable. It is possible that they are referring to the result by A. A. Markov, [117]. I think that the problem is much simpler: cosmologists have failed to note the distinction between a gauge group and a mere symmetry group. In quantum theory, a gauge group is a group of transformations of the algebra of all operators entering the theory, but which leave invariant every hermitian operator representing an observable. On the other hand, a symmetry is an automorphism of the algebra of observables which either commutes with the Hamiltonian, or is part of a dynamical group like the Lorentz boosts. There is at least one observable that is changed by it. If two observers look at the same system, using different coordinates, they see the same world, but obtain different readings on their dials. These must be related by a symmetry transformation, called a passive symmetry by Wigner. Wigner's theorem says that in quantum mechanics, every passive symmetry defines a unitary or anti-unitary operator U on the Hilbert space; this can be used to define an active symmetry: if Φ is a possible state of the system, as seen by an observer O, then so is $U\Phi$, and it can be seen by O. It will be distinguishable from Φ, since U is not a gauge transformation. See [60, 70] for other views on this issue.

In classical field theory, a symmetry shows up as an invariance of the Lagrangian. Thus, the Einstein-Hilbert Lagrangian is invariant under diffeomorphisms of the manifold M, and $Diff\,M$ is a symmetry group of general relativity.

I often hear the argument that, since physics should not depend on the coordinates used, $Diff\,M$ must be a symmetry. This is a fallacious argument; in a coordinate-free formulation of field theory on a manifold, one can choose a dynamics other than that given by the Einstein-Hilbert Lagrangian, and $Diff\,M$ might no longer be a symmetry. Indeed, in any universe with massive particles (like our own), the forces other than gravitational do not possess $Diff\,M$ as a symmetry group. This can easily be seen; such a dynamics is not even invariant under constant scaling of space-time.

In any theory with locality, it does not make sense to regard $Diff\,M$ as a gauge group. An active diffeomorphism might move a state Φ localised near the observer to one far away, and such a transformed state $U\Phi$ is clearly not the same as Φ. It is no good saying that the two states are 'physically equivalent'. This might mean that the diffeomorphed observer, moving with the state, and told to apply the inverse diffeomorphism to his coordinates, would get the same readings from $U\Phi$ as he would have done at his former position on measuring Φ in his former coordinate system. This bewildered observer might mistake Φ for $U\Phi$. Whether they are different states is determined physically: is there any observable which has a different mean in the two states? In statistical physics, the question is, do they both make separate contributions to the partition function (not contradicted by the possibility that they contribute the same amount)? They do, in any theory in which the concept of local observable is meaningful. Observable thermodynamic properties like specific heats, rate of thermalisation *etc.* depend on the multiplicity of the energy-spectrum, and states related by a geometric symmetry such as rotation have the same energy but are not the same state in general. Since statistical mechanics is one of the main tools of cosmology, practitioners of cosmology need to be sure about multiplicity, as measured by a single observer. And there, taking the Ising model as example, the partition function *does* get separate contributions from the various terms that are related to each other by a change in coordinates.

12.9 Dirac Notation

In this part, we discuss the definition of the phase, the operator conjugate to the number operator. It is unwise to embark on a study of the phase operator, if the only tools known to you are those in Dirac's book. For, we quickly arrive at the following paradox.

Theorem 2. *Let h be any positive number; then $h = 0$.*

"PROOF". Consider the self-adjoint operator N defined on the Hilbert space of square-integrable functions on $[0, 2\pi]$; N can be defined by Stone's theorem as the generator of the continuous group of translations

$$U(a/h)\psi(x) := \psi(x - a \,(\text{mod } 2\pi)).$$

Then if the wave-function is differentiable, with square integrable derivative, we have

$$N = (h/i)(d/dx).$$

Thus, the number operator is well defined; its eigenfunctions are $\exp(inx)$, having integer eigenvalues $n = \ldots -1, 0, 1, 2 \ldots$, and are well known to form a complete normalisable orthogonal system. In particular, the constant function $\psi_0 = 1$ is an eigenfunction with eigenvalue zero, also known as the ground state:

$$N\psi_0 = 0.$$

Now define the phase operator X as the operator of multiplication by x:

$$(X\psi)(x) := x\psi(x).$$

Then X is a bounded, everywhere defined operator, and on the dense set of differentiable functions we have the Heisenberg commutation relations

$$[N, X] = h/i,$$

since if $\psi(x)$ is differentiable, so is $x\psi(x)$. Now compute the matrix element of this commutator in the state ψ_0; this is zero, as N hits its ground state either to the left or to the right:

$$0 = \langle\psi_0|NX|\psi_0\rangle - \langle\psi_0|XN|\psi_0\rangle = \langle\psi_0|[N, X]|\psi_0\rangle = h/i\langle\psi_0|\psi_0\rangle = 2\pi h/i.$$

Therefore h, any positive number, must be zero. Think about it. A hint where to find the mistake in the argument will soon be given.

A related paradox is seen from the uncertainty relations; if $qp - pq = ihI$ on a dense domain, then the product of the standard deviations of p and q is greater than $h/2$ (or equal to it for a special wavefunction). So if N, X obey Heisenberg's commutation relations then there can be no eigenstates of N, since for it the variance of X is finite; this is contradicted by the eigenstates $\exp(inx)$, and the fact that X, the phase, is bounded. Where is the mistake in the book proof of the uncertainty relations, when applied to N, X?

I promised to explain the mistake. It occurs because the definition of the differential operator $p = -i\,d/dx$ cannot be extended to all differentiable functions on the circle in a manner which maintains the hermiticity of the operator. Thus the domain includes all differentiable periodic functions, $f(x)$. Then the boundary condition ensures that on integration by parts the boundary terms cancel, and the operator is hermitian. If $f(x)$ is differentiable and periodic, then $xf(x)$ is differentiable, but not always periodic. So the extension of p to include $xf(x)$ does not lead to a hermitian operator. Dirac's bra and ket notation does not mention domains, and so the error is difficult to spot in this notation.

13

Some Good Ideas

"I have started, so I'll finish".
Magnus Magnusson

13.1 Euclidean Quantum Field Theory

Consider the Euclidean version of quantum field theory. This had been introduced first by Schwinger [145] and then by Symanzik [163]. Nelson introduced
the rigorous version [124]. It follows easily from Wightman's version that the
expectation values of a scalar field are analytic in the space-time variables
if we continue from the physical region to the region where time is purely
imaginary. What Nelson noticed and proved for the free field was that there,
the functions define the moments of a generalised classical stochastic process;
the Lorentz invariance of the Wightman functions gives Euclidean invariance
in four dimensions of this process. This was generalised by Glimm and Jaffe
[72] to the interacting field in 2 and in one case, 3, space-time dimensions.
However, no progress was made in solving the problem in four dimensions,
except for the free fields. Two approaches remain; the first is that in quantum
electrodynamics, the Euclidean field exists, but that not all the components
of the four-vector are square-integrable; only the two components leading to
transverse spin are, and so the unphysical components (with negative scalar
products) cancel out the longitudinal components, and we are left with the
free field. This idea has been generalised by Albeverio and Gottfried [1] ; they
construct a Euclidean representation of the field, but not all the states in
Minkowski space have positive probability. This work allows the S-matrix to
be calculated, and so might lead to a possible theory.

13.2 The Most General Quantum Theory

We took quantum mechanics to be based on the algebra $\mathcal{B}(\mathcal{H})$, the set of all bounded operators on the Hilbert space $\mathcal{H}$. This was the situation if there were no superselection rules, and is enough for the study of the Schrödinger equation, for example, of a particle in an external potential. If there is a superselection rule with discrete spectrum, then the observables do not generate an irreducible algebra, and this formalism is not adequate. It is then better to take the Hilbert space to be the direct sum over the values assumed by the superselecting operator, and one may take the algebra of observables to be the direct sum of the algebras, one for each value. One may do the same if Wightman's hypothesis of commuting superselecting rules holds; the corresponding operators then commute and can be simultaneously diagonalised. Even if the spectrum is not discrete, one can form the direct sum over the set of simultaneous values. This leads to a reducible operator algebra as the algebra of observables, on a possible non-separable Hilbert space. In 1964, Haag and Kastler [82] asked how such a scheme comes about; this is the subject of Haag's book on quantum field theory [81].

I.E. Segal concluded in 1947 that in a general setting, any quantum probability should involve a unital C^*-algebra; this is a normed vector space $\mathcal{M}$ with a conjugation $A \mapsto A^*$ and an associative product obeying Gelfand's identity $\|A^*A\| = \|A\|^2$. $\mathcal{M}$ is to contain the identity; it is also complete, a condition which is always true if the algebra is of finite dimension. The self-adjoint elements of $\mathcal{M}$ are the observables, which generalise the concept of random variable. One can add a further property, that $\mathcal{M}$ is generated by its self-adjoint projections, so that there are enough questions. It happens that $\mathcal{B}(\mathcal{H})$ is such a C^*−algebra. Segal considered using a smaller algebra than this; he studied that generated by families of canonical commutation relations. The states one considers can then be limited to *normal* states, those that are countably additive on sums of orthogonal projections. If one likes, one can further limit the states to have finite energy and entropy.

Haag went further than Segal; he limited the algebra to be that generated by fields localised in a finite region of space-time. He had the insight to see that the algebras corresponding to different values of the superselection operators are algebraically isomorphic, but are given by different representations of the abstract algebra. This provides a natural reason for the occurrence of superselection rules: the algebra of the observables has a series of physically realisable representations, each irreducible and inequivalent from each other. This means the following.

Given the C^*−algebra $\mathcal{M}$, a representation of $\mathcal{M}$ is a linear map π from $\mathcal{M}$ into the bounded operators on a Hilbert space $\mathcal{H}$, such that $\pi(A)\pi(B) = \pi(AB)$ for all operators $A, B \in \mathcal{M}$, and $\pi(A)^* = \pi(A^*)$ for all $A \in \mathcal{M}$. One can prove that such a map π is automatically continuous. This follows from the Gelfand relation for the norm on $\mathcal{M}$, and the fact that $\mathcal{M}$ is complete. Two representations π_1 and π_2 of $\mathcal{M}$ on Hilbert spaces $\mathcal{H}$ and $\mathcal{K}$ are said to be equivalent if there exists a unitary operator U from $\mathcal{H}$ onto $\mathcal{K}$, which intertwines the two representations:

$$U\pi_1(A) = \pi_2(A)U \tag{13.1}$$

should hold when applied to any vector in $\mathcal{H}$. We say that a representation π is irreducible if $\mathcal{H}$ has no invariant subspaces, under the action of $\pi(A), A \in \mathcal{M}$ other than the zero vector or the whole space $\mathcal{H}$. One can prove that any C^*-algebra possesses at least one irreducible representation. For the algebras used in quantum field theory, there are in fact infinitely many inequivalent representations. The vectors in the representation spaces then could define possible states of the system. We should, however, limit the representations somewhat; there are representations in which the energy is infinite for all states, and these could not be made in the lab. This question is determined by the *dynamics* of the system, and so we now turn to this.

An *automorphism* of a C^*-algebra $\mathcal{M}$ is a linear map $\tau : \mathcal{M} \to \mathcal{M}$ which preserves products and conjugations, and is invertible. Thus,

$$\tau(A)\tau(B) = \tau(AB)$$
$$\tau(A^*) = \tau(A)^*$$

τ is bijective.

As an example of an automorphism, consider the transformation given by conjugation with a fixed unitary element U of $\mathcal{M}$, thus:

$$\tau(A) = UAU^{-1}, \text{ as } A \text{ runs over } \mathcal{M}. \tag{13.2}$$

Such an automorphism is said to be *inner*. It can be shown that if $\mathcal{M}$ has finite dimension, then every automorphism is inner. More generally, consider an automorphism τ of the (abstract) C^*-algebra $\mathcal{M}$, and a representation π of $\mathcal{M}$ on a Hilbert space, which we denote by $\mathcal{H}_\pi$. We say that τ is *implemented in the representation* π if there exists U, a unitary on $\mathcal{H}_\pi$, such that for each $A \in \mathcal{M}$ we have that Eq. (13.2) holds. We also say that τ is *spatial* in π. Note that here in general the U implementing τ might not be in $\mathcal{M}$, and the automorphism is thus not inner.

A *dynamics* of a C^*-algebra is determined by a group of automorphisms, τ_t, of $\mathcal{M}$, where $t \in \mathbf{R}$, and which obey the additive law:

$$\tau_s \circ \tau_t = \tau_{s+t}. \tag{13.3}$$

Thus, if $A \in \mathcal{M}$ is some observable, we are told that $A(t) := \tau_t(A) \in \mathcal{M}$ is interpreted as that observable at t seconds later. For example, if the Hamiltonian of the system is a bounded operator H in $\mathcal{M}$, then the time evolution is given by

$$\tau_t(A) := A(t) = \exp(iHt)A\exp(-iHt); \tag{13.4}$$

this lies in $\mathcal{M}$ because the series for exp converges, and $\mathcal{M}$ is complete. We say that the automorphism group τ_t is implemented by the unitary operators

$U(t) := \exp itH$. The generator of this automorphism group is defined to be the operator H:

$$\frac{1}{i}\frac{dU(t)}{dt}\big|_{t=0} = HU(0) = H,$$

which can be identified as the energy-operator of the theory. This is not all that useful, since most energy-operators are not bounded. Now, infinite-dimensional C^{*}−algebras possess automorphisms which are not inner, and which are not implemented in all representations. This must occur for the time evolution in a realistic quantum theory. Let us postulate, then, that there is one representation π of the algebra $\mathcal{M}$, with given dynamics τ_t, such that

1. π is irreducible on its separable Hilbert space $\mathcal{H}$;
2. for each time $t \in \mathbf{R}$, τ_t is implemented by $V(t)$ on $\mathcal{H}$;
3. the transition probability $|\langle \Phi, V(t)\Psi \rangle|^2$ is a continuous function of time t.

Item (3) leads us to Wigner's starting point, and allows us to replace the unitaries $V(t)$ by equivalent ones $U(t) = \alpha(t)V(t)$, where $\alpha(t)$ is a real function of time, such that $U(t)$ represent the group of time translations:

$$U(s)U(t) = U(s+t). \tag{13.5}$$

We express the stability of the theory by the postulate that the generator H of this one-parameter group is bounded below. Often, the representation π is just the vacuum representation of the quantum field theory; that is, $\mathcal{H}$ contains a vector Ψ_0 that is invariant under space-time translations, and indeed also under the action of the whole Lorentz group.

When the dynamics is supplemented by the translations, we get an automorphism group acting on $\mathcal{M}$, $\tau_a : a \in \mathbf{R}^4$, and we assume that there is a representation, π, say the vacuum representation, in which each τ_a is implemented by commuting $U(a)$, which obey the group law, and for which the vacuum state is invariant; we then postulate that the simultaneous spectrum of the four-momentum P^μ, $\mu = (0, 1, 2, 3)$, lies in the forward cone.

We seek other representations, not equivalent to π, in which each one-parameter group of space- or time-translations is implemented, and the energy is bounded below. Such representations will be labelled by a superselection quantum number.

13.3 Cohomology

The second idea is due to Haag, and is the following; the dynamics of the field can be expressed as an algebraic dynamics $\tau(t)$ of the local algebra; can we find a new representation of the algebra $\mathcal{M}$, in which each of the automorphisms given by the space-time translations is implemented? Haag proposed that an automorphism, σ, of $\mathcal{M}$, not given by a unitary transformation, can be used as follows. Define the representation $\pi_\sigma(A)$ of the operator $A \in \mathcal{M}$ by

$$\pi_\sigma(A) = \pi(\sigma(A)). \tag{13.6}$$

We see that this is an operator on the original Hilbert space $\mathcal{H}$. Indeed, the map $A \mapsto \pi_\sigma(A)$ is a representation of $\mathcal{M}$. We have to check the linearity of the map, and also that it preserves the multiplication law of the algebra, and also the $*$−conjugation. Thus, for linearity, we see that

$$\pi_\sigma(A + B) = \pi(\sigma(A + B)) = \pi(\sigma(A) + \sigma(B))$$

by the linearity of σ

$$= \pi(\sigma(A)) + \pi(\sigma(B))$$

by the linearity of the representation π

$$= \pi_\sigma(A) + \pi_\sigma(B).$$

To show that $\pi_\sigma(AB) = \pi_\sigma(A)\pi_\sigma(B)$, we similarly use that $\sigma(AB) = \sigma(A)\sigma(B)$ and that $\pi(AB) = \pi(A)\pi(B)$. That $\pi_\sigma(A^*) = (\pi_\sigma(A))^*$ is similarly proved. Thus, π_σ is a $*$−representation of the C^*−algebra $\mathcal{M}$. Note that this represents an operator on $\mathcal{M}$ by some operator on the original Hilbert space $\mathcal{H}$.

Suppose that, contrary to Haag's assumption, σ is spatial in the representation π; then there is a unitary operator W on $\mathcal{H}$ such that

$$\sigma(A) = WAW^{-1}$$

for all elements $A \in \mathcal{M}$. In that case, we have

$$\pi(\sigma(A)) = W\pi(A)W^{-1}$$

holds for all $A \in \mathcal{M}$. We then see from Eq. (13.6) that

$$\pi_\sigma(A) = \pi(\sigma(A)) = W\pi(A)W^{-1}$$

also holds for all $A \in \mathcal{M}$; this states that π_σ is equivalent to π; this explains why Haag needed to choose σ so that it is NOT SPATIAL in the representation π. If this is the case, and we impose the requirement on σ that each space-time automorphism should be implemented in π, then we can get some dynamics, by using Wigner's methods.

Haag's assumption has been generalised by Doplicher and Roberts, that we impose an endomorphism, not just an automorphism, to change the representation. Recall that an endomorphism σ of an algebra obeys the linearity and multiplicative axioms of an automorphism, and also preserves the $*$−operation for a C^*−algebra. The only difference with an automorphism is that the mapping may not be invertible. However, this last property was not used above, in our proof that π_σ defines a representation. Thus, we may use an endomorphism instead of an automorphism to get our new representation of the observables. These authors have examined the possibilities, and do not come to any conclusion in four dimensional space-time. I have suggested that one can relax the condition that the representation of the space-time translations should be true; if we allow a projective representation, then it might be possible to find examples, of the free field in one representation giving rise to interaction in another. There will be no direct interaction between photons; it would need to come from producing other particles in its intermediate states. These would be of finite energy, but would contain infinitely many photons. This idea has been successful in two-dimensional theories [161].

Let us start with free quantum electrodynamics. This is related to the free quantum field, obtained by second-quantising the irreducible representation of $\mathcal{P}_+^\uparrow$, the inhomogeneous Lorentz group. The Hilbert space of this field is the Fock space over the one-particle space; this space carries the direct sum of two irreducible representations of the Poincaré group, $\mathcal{P}_+^\uparrow$; these two representations are related by parity. The free quantised field gives a reducible representation of $\mathcal{P}_+^\uparrow$, determined by the direct sum of the vacuum, the one-particle photon, the two-particle photons, and so on. The free field is generated by the spectral projections of the self-adjoint operators $\int \mathbf{E}(x)f(x)\,d^4x$ and $\int \mathbf{H}(x)g(x)\,d^4x$. Here, f and g are smooth functions of their variables, and have compact support. Let us now use the idea of Doplicher and Roberts that we should look for the representations of the algebra of observables that is obtained from a given representation by an endomorphism of the observables. This will lead to a reducible representation of the observable algebra. We seek interesting representations of the algebra; these must possess positive energy. This would be defined as the generator of the unitary group implementing the time-evolution group. It can be shown that any one-parameter group of unitaries, up to a multiple of the identity, can be replaced by a true representation. However, this result holds only for each one-parameter subgroup of the group $\mathbf{R}^4$ of space-time translations; it does not apply to $\mathbf{R}^4$. The representation of the inhomogeneous Lorentz group is reducible, so Wigner's theorem does not apply. Recall that Wigner showed that any irreducible projective representation of $\mathcal{P}_+^\uparrow$ can be reduced to a true representation of the covering group. This is the inhomogeneous $SL(2,\mathbf{C})$, and has the usual integer and half-integer representations of mass $m > 0$ and the usual helicity representations of mass equal to 0. However, in our case it remains to be proved that we cannot have a multiplier that is an element in the commutant, which cannot be removed to a scalar multiple of the representation. This is true even if we only consider the abelian group $\mathbf{R}^4$ as the group in question. We would then get a non-abelian representation of the space-time translation group, which leads us into cohomology theory [156].

Let us deal with these questions. We start with the free second-quantised electromagnetic field, there being no rigorous interacting model we could have chosen. Choose space coordinates so that c, the speed of light, is unity. Let $\mathbf{E}$ and $\mathbf{H}$ be the electric and magnetic fields. These together form a tensor; let μ and ν run over the indices $0, 1, 2, 3$. Then

$$F_{\mu\nu}(x) := \begin{pmatrix} 0 & E_1 & E_2 & E_3 \\ -E_1 & 0 & H_3 & -H_2 \\ -E_2 & -H_3 & 0 & H_1 \\ -E_3 & H_2 & -H_1 & 0 \end{pmatrix}. \tag{13.7}$$

It is known that $\int d^4x F_{\mu\nu}(x) f(x)$ is self-adjoint whatever the values of μ and ν, for all f of compact support and infinitely differentiable; that is, for all $f \in \mathcal{D}(\mathbf{R}^4)$. We can construct the C^*−algebra $\mathcal{M}$ from the spectral resolutions of all these field components. Recall that the algebra is the norm closure of the algebra of operators generated by these spectral resolutions; the fields themselves are unbounded operators, and so do not lie in our $\mathcal{M}$. The space-translation group acts on $\mathcal{M}$, since it acts on $\mathcal{D}$. The time-evolution also acts on $\mathcal{M}$; we just need to follow the free Maxwell's equations from the time zero to any other time; the fields transform linearly, and the test functions transform to other test functions back at time zero. The Lorentz group also acts on the space of tensors. It is clear that the action of $\mathcal{P}_+^\uparrow$ on $\mathcal{A}$ is given by unitary operators, which provide a representation of this group. Indeed, the unitary operators which implement (a, Λ) are just the second quantisation of the representation $(0, 1, +) \oplus (0, 1, -)$. Let U_0 denote this action; for example, for $a = (\mathbf{a}, a_0)$ in the space-time translation group, we get

$$U_0^{-1}(a) F_{\mu\nu}(\mathbf{x}, t) U_0(a) = F_{\mu\nu}(\mathbf{x} + \mathbf{a}, t + a_0). \tag{13.8}$$

So much for the free field. The representation $\pi(A) = A$ then obeys the Haag axioms.

We get some other representations of $\mathcal{M}$ by applying a ∗-endomorphism to it. Recall that a ∗-endomorphism of $\mathcal{M}$ is a linear map σ on $\mathcal{M}$ such that

1. $\sigma I = I$;
2. $\sigma(AB) = \sigma(A)\sigma(B)$ for all $A, B \in \mathcal{M}$;
3. $\sigma(A^*) = (\sigma(A))^*$ for all $A \in \mathcal{M}$

for all $A, B \in \mathcal{M}$. This is more general than an automorphism, since the latter has to be invertible. We checked already that we may represent the element $A \in \mathcal{A}$ by the operator σA, acting on the original space. So we have a representation π_σ, of our algebra $\mathcal{M}$, given by $\pi_\sigma(A) = \sigma A$. It will not in general be equivalent to the free representation. For example, it might be reducible, if M has no inverse. I suggest that such a representation might incorporate interacting particles. This differs from Haag's idea; he would have needed $\mathcal{M}$ to be the interacting field, not the free field as here.

In order for the lab to be able to produce the representation π_σ, its energy-momentum would need to be finite. However, it is not obvious that the space-time automorphisms of $\mathcal{M}$ are given by unitary operators in the representation π_σ; we do not want space-time symmetries to be *spontaneously broken*. This occurs when a symmetry of the *algebraic* dynamics is not given by a unitary operator. This could happen if σ does not commute with space-time translations, which we must allow. Thus, we require for each space-time translation $a = a_\mu$ a unitary operator, say $U(a)$, such that

$$U(a)\pi_\sigma(A)U^{-1}(a) = \pi_\sigma(A_a) \tag{13.9}$$

for all $A \in \mathcal{M}$. Using the definition, this gives

$$U(a)\sigma A U^{-1}(a) = \sigma A_a = \sigma\left(U_0(a)AU_0^{-1}(a)\right) \tag{13.10}$$

for all A. To find U is not so easy; recall that σ does not commute with U_0. It might not exist at all. If one such $U(a)$ exists, then it is not unique; for, we may replace it with $uU(a)v$ for any $u, v \in (\sigma\mathcal{M})'$, the set of operators that commute with $\sigma\mathcal{M}$. Thus, we see that the unitary operators have become gauge dependent.

Suppose that we have chosen σ well, so that we can get $U(a)$ with the property (13.10), and that this can be done for any $a \in \mathbf{R}^4$. We then find that $\omega := U(a+b)^{-1}U(a)U(b)$ commutes with any element of $(\sigma\mathcal{M})$: just put the whole thing into (13.10). But unlike Wigner's case, when by Schur's lemma it must be a multiple of the identity, all we can say is that $\omega \in (\sigma\mathcal{M})'$. So instead of the group law, we get the multiplier relation

$$U(a)U(b) = \omega(a,b)U(a+b). \tag{13.11}$$

Indeed, there is one relation that is obeyed by ω: it is a 2-cocycle. This follows from the associativity of multiplication in the group $\mathbf{R}^4$. By evaluating $a + (b+c) = (a+b)+c$, $a, b, c \in \mathbf{R}^4$, and using (13.11), we see that

$$(U(a)U(b))U(c) = \omega(a,b)U(a+b)U(c) = \omega(a,b)\omega(a+b,c)U(a+b+c)$$
$$U(a)(U(b)U(c)) = U(a)\omega(b,c)U(b+c) = \gamma_a(\omega(b,c))U(a)U(b+c)$$
$$= \gamma_a(\omega(b,c))\omega(a,b+c)U(a+b+c) \tag{13.12}$$
$$\text{where } \gamma_a(A) := U(a)AU^{-1}(a).$$

Thus we get the relation

$$\omega(a,b)\omega(a+b,c) = \gamma_a(\omega(b,c))\omega(a,b+c). \tag{13.13}$$

This is the cocycle relation.

As an example, consider the identity endomorphism. Then π_ι is equivalent to the free representation, and $U(a)$ exists, but differs from $U_0(a)$ by a phase, $u(a)$. Thus, for each $a \in \mathbf{R}^4$, we get $U(a) = u(a)U_0(a)$. Since $U_0(a)U_0(b) = U_0(a+b)$ for all $a, b \in \mathbf{R}^4$, and complex numbers commute with each other and with $U_0(a)$, we see that in this case, we have

$$U(a)U(b)U(a+b)^{-1} = u(a)U_0(a)u(b)U_0(b)U_0(a+b)^{-1}u(a+b)^{-1}$$

so in this case

$$\omega(a,b) = u(a)u(b)u(a+b)^{-1}. \tag{13.14}$$

One can check that this satisfies the cocycle relation, as it must. We say that a cocycle of the form (13.14) is a *coboundary*. It arises because the unitary operators implementing the space-time group are ambiguous; this clearly has no physical meaning, so we agree to use the ambiguity to reduce the ω to the identity when this is possible, by premultiplying $U(a)$ by $u^{-1}(a)$ so that U becomes equal to $U_0(a)$. It will not be possible, for the general cocycle, to reduce it to $U_0(a)$, or to get a true representation from $U(a)$ by gauge transformations.

We could also require that each element of the Lorentz group should be implemented by a unitary operator in the representation π_σ; this would ensure that they are still symmetries in the new representation. However, even without this, we maintain the locality. That is, if A and B are local observables that are space-like separated, then $[A, B] = 0$; we get

$$[\pi_\sigma(A), \pi_\sigma(B)] = [\sigma A, \sigma B] = \sigma A \sigma B - \sigma B \sigma A = \sigma(AB) - \sigma(BA)$$
$$= \sigma(AB - BA) = 0$$

Thus, most if not all the properties of Lorentz invariance will be obtained from the representation π_σ. To find such an operator σ, and look for the consequences, would be an interesting enterprise. Indeed, it can be proved that under similar circumstances when the cocycle is a coboundary that the scattering matrix is Lorentz invariant.

The resulting theory would have a gauge group acting, and the space-time translations would not commute with each other. If σ is an endomorphism, but is not an automorphism, then the gauge group would be non-abelian. It might work.

References

1. Albeverio S., and Gottschalk H., *Scattering theory for quantum fields with indefinite metric*, Commun. Math. Phys., **216**, 491–513, 2001
2. American Academy of Pediatrics, http://pediatrics.aappublications.org/cgi/content/full/100/5/735
3. Amelino-Camelia G., *Relativity in space-times with short-distance structure governed by an observer-independent (Planckian) length-scale*, Intern. J. Modern Phys., **D11**, 35-, 2002
4. Andréasson S. et al., *LANCET*, 1987, (ii), 1483-5
5. Araki H., Prog. Theoret. Phys., **64**, 719–730, 1980
6. Araki H., *A continuous superselection rule as a model of classical measuring apparatus in quantum mechanics*, pp 23–33 in **Fundamental Aspects of Quantum Theory**, eds. V. Gorini and A. Frigerio, NATO ASI series **144**, Plenum Press, 1986
7. Araki H., **Mathematical Theory of Quantum Fields**, Oxford Univ. Press, 2000
8. Arsenault L. et al., Brit. Med. J, Vol **325**, 1212–1213, 2002
9. Arsenault L. et al., The British Journ. of Psychiatry, Vol **184**, 110–117, 2004
10. Aspect A., Grangier P. and Roger G., *Experimental tests of realistic local theories via Bell's theorem*, Phys. Rev. Lett., **47**, 460–463, 1981
11. Aspect A., Grangier P. and Roger, G., *Experimental realization of EPR-Bohm gedankenexperiment. A new violation of Bell's inequalities* Phys. Rev. Lett., **49**, 91–94, 1982
12. Aspect A., Dalibrand J. and Roger, G., *Experimental test of Bell's inequalities using time-varying analysers*, Phys. Rev. Lett., **49**, 1804–1807, 1982
13. Bargmann V., *On unitary ray representations of continuous groups*, Annals of Math., **59**, 1–46, 1954
14. Barndorff-Nielson O. E., Gill R. D. and Jupp P. E., *On quantum statistical inference*, J. Royal Stat. Soc., **B65**, 775–816, 2003
15. Barton G. and Dare D., *Some implications of the bootstrap model for the EM properties of baryons*, Phys. Rev. **150**, 1220–1231, 1966
16. Bass L., *How to predict everything: Nostradamus in the role of Copernicus*, Reports on Mathematical Phys., **57**, 13–15, 2006
17. Bell J. S., *On the Einstein-Podolski-Rosen paradox*, Physics, **1**, 195–200, 1964
18. Bell J. S., Helv. Phys. Acta, **48**, 80-, 1975

19. Bell J. S., **The Speakable and Unspeakable in Quantum Mechanics**, Camb. Univ. Press, 1987.
20. Bell, Stuart, **When Salem Came to the Boro**, Pan, 1988
21. Bender C. M., Brody D. and Jones H. F., *Complex extension of quantum mechanics*, Phys Rev Lett **89**, 270402, 2002
22. Bender C. M., Brody D. C. and Jones H. F., *Scalar quantum field theory with a complex cubic interaction*, Phys. Rev. Lett., **93**, 251601, 2004
23. Bohm D., **Quantum Theory**, Prentice-Hall, New York, 1951
24. Bohm D., *A suggested interpretation of quantum theory in terms of hidden variables*, I, II. Phys. Rev., **85**, 166–179 and 180–193, 1952.
25. Bohm D., **Causality and Chance in Modern Physics**, Routledge Kegan Paul, London 1957
26. Bohm D., **Wholeness and the Implicate Order**, Routledge Kegan Paul, London, 1980
27. Bohm D. and Vigier J. V., Phys. Rev., **96**, 208-, 1954
28. Bohm D. and Hiley B., **The Undivided Universe**, Foundations of Physics, **25**, 507-, 1995
29. Bohm D. and Hiley B., **The Undivided Universe**, Routledge, London, 1993
30. Bohr N., **Atomic Physics and Human Knowledge**, 81, Wiley, 1958.
31. Bohr N., **Essays 1958–1962 on Atomic Physics and Human Knowledge**, 15, Wiley, 1963
32. Borchers H. J., *Über die Mannigfaltigheit der interpolierenden Felder zu einer kausalen S-matrix*, Nouvo Cimento, **15**, 784-, 1960
33. Bremmermann H. J., Oehme R. and Taylor J. G., *A Proof of Dispersion Relations in Quantum Field Theories*, Physical Review, **109**, 2178-, 1958.
34. Brody D. C., http://theory.ic.ac.uk/~brody/DCB/dcb33.pdf; *Must a Hamiltonian be Hermitian?*
35. Brooks, Michael, *Can quantum quirk give objects mass?* **New Scientist**, 23 Oct. 2004, page 10.
36. Bub J., *von Neumann's projection postulate as a probability conditionalization rule in quantum mechanics*, J. Phil. Logic, **6**, 381–390, 1977
37. Bucholz D., Commun. Math. Phys., **85**, 49-, 1982
38. Bucholz D. and Fredenhagen K., Commun. Math. Phys., **84**, 1-, 1982
39. Caliceti E., Graffi S. and Maioli M., Commun. Math. Phys., **75**, 51-, 1980
40. Castillejo L., Dalitz R. and Dyson F. J., Phys. Rev. **101**, 453-, 1956
41. Caves C., *Predicting future duration from present age: a critical assessment*, Contemporary Physics, **41**, 143–153, 2000
42. Chew G. and Mandelstam S., Phys. Rev. **126**, 1202-, 1962
43. Christensen P. V., http://mmf.ruc.dk/~PVC/joensuu.htm
44. Conte E., Todarello O., Federici A., Vitiello F., Lopane M. and Krennikov A., *Preliminary evidence of quantum-like behaviour in measurements of mental states*, in **Quantum Theory: Reconsideration of Foundations − 2**, Editor, Krennikov, A. Växjö University Press, Växjö, Sweden.
45. Dashen R. F. and Frautschi S. C., *Bootstrap theory of octet enhancement*, Phys. Rev. Lett., **13**, 497–500, 1964
46. Dashen R. F. and Frautschi S. C., Phys. Rev. **135**, B1190-, B1196-, 1964
47. Dashen R. F. and Frautschi S. C., *Weak and electromagnetic interactions of the hadrons in bootstrap theory*, Phys. Rev. **143**, 1171–1184, 1966
48. Davies E. B., **Quantum Theory of Open Systems**, Academic Press, 1978.

49. Davies E. B., **Science in a Looking Glass**, Oxford Univ. Press, 2004
50. Davies E. B. and Lewis J., T., *An operational approach to quantum probability*, Commun. Math. Phys., **17**, 239–260, 1970
51. deWitt B. S. and Graham N, **The many worlds interpretation of quantum mechanics**, Princeton Univ. Press, 1986
52. Dirac P. A. M., **The Principles of Quantum Mechanics**, 4th ed., Clarendon Press, Oxford, 1958
53. Doplicher S., Haag R. and Roberts J., Commun. Math. Phys., **23**, 199-, 1971; **35**, 49-, 1974
54. Doplicher S. and Roberts J., Commun. Math. Phys., **131**, 51-,1990
55. D'Souza C. et al, *Marijuana and Madness*, ed. D. J. Castle and R. Murray, Cambridge University Press, 2004. Degenhardt, L. and Hall, W., *Australian and New Zealand Journal of Psychiatry*, Vol **34**, 26–34, 2004
56. Dürr D., Fusseder W., Goldstein S. and Zanghi N., *Comment on Surrealistic Bohm Trajectories*, Zeits. fur Naturforschung, **48a**, 1261–1262, 1993
57. Durr D., Goldstein S. and Zanghi N., J. Stat. Phys., **67**, 843–907, 1992
58. Dürr D., Goldstein S. and Zanghi N., in **Bohmian Mechanics and Quantum Theory**, eds. Cushing J. T., Fine A. and Goldstein S., Kluwer, 1996
59. Dyson F. J., **Physics Today**, **18**, 21–24, 1965
60. http://www.edge.org/documents/archive/edge12.html
61. Englert B.-G., Scully M. O., Süssmann G. and Walthier H., *Surrealistic Bohm Trajectories*, Z. fur Naturforschung, **47a**, 1175–1286, 1992
62. Einstein A., Podolski B. and Rosen N., *Can quantum-mechanical description of physical reality be considered complete?*, Phys. Rev. **47**, 777–780, 1935
63. Everett H., *Relative state formulation of quantum mechanics*, Rev. Mod. Phys., **29**, 454–462, 1957
64. Feldman M., http://ourworld.compuserve.com/homepages/Marc_Feldman _2/
65. Fényes I., *Eine wahrscheinlichkeitstheoretische Begründung und Interpretation der quantenmechanik*, Zeitschrift für Physik, **132**, 81–106, 1952
66. Friedrichs K. O., Shapiro H. N. et al., **Integration of Functionals**, New York University, Institute of Mathematical Sciences, 1957
67. Froissart M., Nuovo Cimento, **B64**, 241-, 1981
68. Wallace Garden R., **Quantum Theory and Modern Logic**, Adam Hilger, 1984.
69. Ghose P., *An experiment to distinguish between de Broglie-Bohm and standard quantum mechanics*, http://arXiv.org/PS_cache/quant-ph/pdf/0003/0003037.pdf
70. Giulini D., Joos E., Kiefer C., Kupsch J., Stamatescu I.-O. and Zeh H. D., **Decoherence and the Appearance of a Classical World in Quantum Theory**, Springer-Verlag, Berlin, 1996
71. Gleason A., *Measures on the closed subspaces of a Hilbert space*, J. Math. Mech. **6**, 895–894, 1953
72. Glimm J. and Jaffe A. M., **Quantum Physics**, second ed. Springer-Verlag, New York, 1987
73. Goldstein S., J. Stat. Phys., **47**, 645–667, 1987
74. Gott J. R., *Implications of the Copernican principle for our future prospects*, Nature **363**, 315–319, 1993
75. Gott J. R., *Future prospects discussed*, Nature, **368**, 108, 1994

76. Greenberger D. M., Horne M. A. and Zeilinger A., *Going beyond Bell's theorem*, in: **Bell's Theorem, Quantum Theory and Conceptions of the Universe**, ed. M. Kafatos, 1988.

77. Greenberger D. M., Horne M. A., Shimony A. and Zeilinger A., Amer. J. Phys., **58**, 1131–1143, 1990

78. Griffiths R. B., J. Stat. Phys., **36**, 219-, 1984

79. Griffiths R. B., Amer. J. Phys., **55**, 1-, 1987

80. Haag R., Ann. Phys. (Leipzig), **11**, 29-, 1963

81. Haag R., **Local Quantum Physics**, second ed., Springer-Verlag, 1999

82. Haag R. and Kastler D., *An algebraic approach to quantum field theory*, J. Mathematical Phys., **5**, 848–861, 1964

83. Haag R. and Schroer B., *The postulates of quantum field theory*, J. of Mathematical Phys., **3**, 248–256, 1962

84. Hardy L., Phys. Rev. Lett. **68**, 2981–2988, 1992

85. Hardy L. and Squires E., Phys. Lett. **168**, 169-. 1992

86. Heisenberg W., **The Physical Principles of Quantum Theory**, Dover, Chicago, 1930

87. Hepp K., Helv. Phys. Acta., **45**, 237-, 1972

88. Hepp K., Helv. Phys. Acta, **48**, 80-, 1975

89. Hepp K. and Lieb E. H., Helv. Phys. Acta, **46**, 573-., 1974

90. Hill R., *Reflections on the cot-deaths cases*, Significance, 13–15, March 2005

91. Hill R., *Multiple sudden infant deaths – coincidence or beyond coincidence?* Paediatric and Perinatal Epidemology, **18**, 320–326, 2004

92. Lord Howe, http://www.parliament.the-stationery-office.co.uk/pa/ ld199697/ldhansrd/pdvn/lds01/text/11017-08.htm

93. Huynh Kiet T., www.medicine.uiowa.edu/pa/sresrch/huynh/huynh/*now deleted*

94. Jaynes E. T., *Prior probabilities*, IEEE Transactions on Systems Science and Cybernetics, SSC-4, 227–241, 1968

95. Jeffreys H., **Theory of Probability**, Oxford Univ. Press, 1939

96. Jona-Lasinio G., Martinelli F. and Scoppola E., Commun. Math. Phys. **80**, 223-, 1981. Claverlie P. and Jona-Lasinio G., Phys. Rev. A, **33**, 2245-, 1986

97. Jost R., in **Essays in Honour of Valentine Bargmann**, (Eds. Lieb E. H. et al.), Princeton Univ Press, 1976

98. Kant, Immanuel, **Kritik der reinen Vernuft**, Königsberg, 1781

99. Kauffman S. and Smolin L., *A possible solution for the problem of time in quantum cosmology*, http://www.edge.org/3rd_culture/smolin/smolin_p1.html

100. Kendall D. J. and Kendall W. S., *Alignment in two-dimensional sets of points*, Adv. to Appl. Probab., **12**, 360–434, 1980

101. Kent A., *Against many worlds interpretation*, Int J. Mod. Phys., **A5**, 1745–1762, 1990

102. Kierland B. and Monton B., *How to predict future duration from present age*, http://opp.weathersen.net/archives/004350.html, to appear in The Philosophical Quarterly

103. Kochen S. and Specker E. P., *The problem of hidden variables in quantum mechanics*, J. Math. Mech., **17**, 59–87, 1967

104. Kim Y. S., Phys. Rev. **142**, 1150–53, 1966

105. Kim Y. S., http://www2.physics.umd.edu/~yskim/home/dashen.html

106. Kim Y. S., and Vasavada, K. V., Phys. Rev. **150**, 1236–40, 1966. *Erratum*: Phys. Rev. **172**, 1849, 1968
107. Kim Y. S., and Vasavada, K. V., Phys. Rev. **D5**, 1002–1010, 1972
108. Kolmogorov, A. N., **Grundbegriffe der Wahrscheinlichkeitsrechnung**, Moscow, 1933
109. Krylov N. S., **Works on the Foundations of Statistical Physics**, Princeton University Press, 1979
110. Landau J. L., *On the violation of Bell's inequality in quantum theory*, Phys. Lett. **A120**, 54–56, 1987
111. Lovelace C., Proc. Roy. Soc., **A289**, 547-, 1966
112. Lüders G., *Über Zustandsänderung durch den Messprozess*, Ann. der Physik, **8**, 322–328, 1951
113. Mackey G. W., **The theory of unitary group representations**, University of Chicago Press, 1976
114. Madelung E., Z. Physik **40**, 322-, 1926
115. Magueijo J., **Faster Than the Speed of Light: The Story of a Scientific Speculation**, Perseus, 2003; in the UK: Heinemann
116. Markopoulou F. and Smolin L., http://arxiv.org/abs/gr-qc/0311059
117. Markov A. A., *Insolubility of the problem of homeomorphy*, pp 300–306, **Proc. Intern. Congress of Math.**, 1958, Cambridge Univ. Press, Cambridge, 1958
118. Meadow R., **The ABC of Child Abuse**, page 29, BMJ Publ. Group, 1997
119. Mermin N. D., Phys. Rev. Lett., **65**, 3373–3376, 1990
120. M.A.M.A., http://www.mspb.com/
121. Morato L. M. and Petroni N. C., J. Phys A, **33**, 5833–5848, 2000
122. Nelson E., **Dynamical Theories of Brownian Motion**, Princeton Univ. Press, 1967
123. Nelson E., **Quantum Fluctuations**, Princeton Univ. Press, 1985
124. Nelson E., *Taking Formalism Seriously*, The Mathematical Intelligencer, Springer, **15**, #3, 8-, 1993
125. Nelson E., *Field Theory and the Future of Stochastic Mechanics*, in **Stochastic Processes in Classical and Quantum Systems**, Lect. Notes in Phys **262**, p 438, Springer-Verlag, 1986
126. Neumaier A., *Bohmian mechanics contradicts quantum mechanics*, http://arXiv.org/PS_cache/quant-ph/pdf/0001/000101.pdf
127. Neumaier A., *Ensembles and experiments in classical and quantum physics*, Int. J. Mod. Phys., **B17**, 2937–2980, 2003
128. Nott M., Medical expert testimony, *A failure to learn lessons*, Lawyers' Weekly Lexis Nexis, 16 Sept. 2005, Sydney
129. The New Scientist, http://www.newscientist.com/hottopics/quantum/
130. Omnès R., J. Stat. Phys., **53**, 893-, 1988
131. Omnès R., **The Interpretation of Quantum Mechanics**, Princeton University Press, 1994
132. Pagels H. R., Phys. Rev. **144**, 1261–69, 1966
133. Penrose R., **The Emperor's New Mind**, Vintage Books, 1989
134. Penrose R., **Shadows of the Mind**, Oxford University Press, 1994
135. Percival I., **Quantum State Diffusion**, Cambridge University Press, 1998
136. Popper K., **The Logic of Scientific Discovery**, Hutchinson, London, 1959
137. Popper K., **Quantum Theory and the Schism in Physics**, Routledge, London, 1982

138. del Prete, Valeria, in http://www.mth.kcl.ac.uk/staff/v_del_prete.html

139. Prugovecki E., in http://individual.utoronto.ca/prugovecki/ EpistemicPerspectives.html

140. Rauch H., Zeilinger A., Badurek G., Wilfing A., Bauspiess U. and Bonse U., *Verification of a coherent spinor rotation of fermions*, Physics Letters, A54, 425–427, 1975.

141. Ruelle D., , *On the asymptotic condition in quantum field theory*, Helv. Phys. Acta, **35**, 147-, 1962

142. Ruelle D., **Statistical Mechanics; Rigorous Results**, W. A. Benjamin, New York, 1969

143. Sawyer R. F., Phys. Rev. **142**, 991–995, 1966

144. Schroer B., in http://arxiv.org/abs/hep-th/0303241

145. Schwinger J., *On the theory of quantized fields, I*, Phys. Rev. **82**, 664- and 914-, 1951

146. Segal I. E., , J. Mathematical Phys., **5**, 269-, 1964. Bull. Proc. Math. France, **91**, 129-, 1963.

147. Segal I. E., **Mathematical Problems of Relativistic Physics**, Amer. Math. Soc., 1963

148. Siegel A. and Wiener N., *'Theory of measurement' in differential space quantum theory*, Phys. Rev. **101**, 429–432, 1956

149. Small C. G., **The Statistical Theory of Shape**, Springer-Verlag, New York, 1996

150. Souriau J.-B., **Structures des systemes dynamiques**, Dunod, 1970

151. Southall D. P., Plunkett M. C., Bank M. W., Falkov A. F. and Samuels M. P., *Covert video recordings of life-threatening child abuse*, Pediatrics, **100**, 735–760, 1997 Available at http://pediatrics.aappublications.org/cgi/content/full/100/5/735

152. Stapp H. P., **Mind, Matter and Quantum Mechanics**, Springer-Verlag, Heidelberg, Second ed., 2004

153. Stapp H. P., http://psyche.cs.monash.edu.au/v2/psyche-2-05-stapp.html

154. Streater R. F., *Canonical quantisation*, Commun. Math. Phys., **2**, 354–374, 1966

155. Streater R. F., *The representations of the oscillator group*, Commun. Math. Phys., **4**, 217–236, 1967

156. Streater R. F., *Symmetry groups and non-abelian cohomology*, Commun. Math. Phys., **132**, 201–215, 1990.

157. Streater R. F., **Statistical Dynamics**, Imperial College Press, 1995.

158. Streater R. F., *Stability of a hot Smoluchowski fluid*, Open Systems and Info. Dynamics, **7**, 1–9, 2000

159. Streater R. F. and Wightman A. S., **PCT, Spin and Statistics, and All That**, Benjamin, 1964. Third Edition, Princeton University Press, 2000

160. Streater R. F., http://www.maths.kcl.ac.uk/~streater/lostcauses#XI.html

161. Streater R. F. and Wilde I. F., *Fermion States of a Boson Field*, Nucl. Phys., **B24**, 561–575, 1970

162. Strocchi F. and Wightman A. S., J. Math. Phys. **15**, 2198-, 1974; **17**, 1930-, 1976

163. Symanzik K., pp 152–226 in *Local Quantum Theory*, ed. R. Jost, Academic Press, 1969

164. Tsirelson B. S., *Quantum analogues of Bell's inequalities* (Russian), Zap. Nauchn. Sem. Leningrad Odtel. Mat. inst. Steklov (LOMI), **142**, 174–194, 1985

165. Tsirelson B. S., *Some results and problems on quantum Bell-type inequalities*, Hadronic Journal Suppl., **8**, 329–345, 1993

166. von Neumann, J., **Die mathematischer Grundlagen der Quantenmechanik**, Budapest, 1932; translated as **The Mathematical Foundations of Quantum Mechanics**, Princeton University Press, 1955

167. Wick G. C., Wightman A. S. and Wigner E. P., Phys. Rev. **88**, 101-, 1952

168. Wiener N. and Siegel A., *A new form for the statistical postulate of quantum mechanics*, Phys. Rev. **91**, 1551–1560, 1953

169. Wiener N. and Siegel A., *The differential-space theory of quantum systems*, Nuovo Cimento, **2**, Ser X (Supp No 4), 982–1003, 1955

170. Wiener N., Siegel A., Rankin B. and Martin W., **Differential Space, Quantum Systems and Prediction**, M.I.T. Press, Cambridge, Mass., 1966

171. Wightman A. S., *Superselection Rules: Old and New*, Nuovo Cimento, **B110**, 751–769, 1995

172. Wigner E. P., **Gruppentheorie und ihre Anwendung auf die Quantenmechanik der Atomspectrum**, Friedr. Vieweg Braunschweig, 1931; English translation: **Group theory and its application to the quantum mechanics of Atomic Spectra**, Academic Press, 1959

173. Wollenberg L., Ph. D. thesis, Oxford, 1973

Index